Computer
Telephony
Integration

Computer Telephony Integration

William A. Yarberry, Jr.

Boca Raton   London   New York   Washington, D.C.

Library of Congress Cataloging-in-Publication Data

Yarberry, William.
    Computer telephony integration / William Yarberry, Jr.
        p. cm.
    Includes bibliographical references and index.
    ISBN 0-8493-9995-5 (alk, paper)
    1. Telematics. 2. Internet telephony. 3. Digital telephone systems. I. Title.
TK5105.6.Y37   1999
621.385–dc21
                                                               99-31620
                                                                          CIP

No claim to original U.S. Government works
International Standard Book Number 0-8493-9995-5
Library of Congress Card Number 99-31620
Printed in the United States of America 1 2 3 4 5 6 7 8 9 0
Printed on acid-free paper

Dedication

To Carol, Will, Libby, and my parents

Contents

Preface

THERE ARE MANY TELEPHONY BOOKS THAT PROVIDE DETAILED INSTRUC-
TIONS TO THE READER on *how* to assemble and troubleshoot voice communications systems. This book, in contrast, is targeted at the supervisor,
manager, or director that must decide *what* to do. Communications demands in the 21st century are intense. Computing increasingly dominates
the telephony world. *Computer Telephony Integration* addresses many of the
day-to-day technologies encountered by organizations that are merging
their voice and data infrastructures. The relentless increase in digital signal
processing power over the next few years will enable fabulous new applications in voice recognition technology, seamless conversion of content from
one media form to another, and intelligent routing of calls. Desktops and applications of all types will be increasingly telephony enabled using tools
such as ActiveX or JTAPI with an Internet browser. Call centers are becoming smaller and more geographically dispersed. And, of course, the big
"kahuna" of telecommunications technologies — Voice over IP — is poised
to completely change the landscape of voice communications.

The typical communications manager faces not only rapid changes in
technology and standards, but also must control costs, develop RFPs, install new PBXs, provide call accounting, and maintain adequate security.
These mundane but critical functions are discussed alongside the more ex-
citing technology developments in order to present a balanced view of telephony management concerns.

Reader comments and questions are most welcome. Please send them
via e-mail to the author at azuresky@hal-pc.org.

Acknowledgments

MANY INDIVIDUALS HAVE CONTRIBUTED BOTH DIRECTLY AND INDIRECTLY TO THIS BOOK. My editor, Christian Kirkpatrick, has provided encouragement, advice, and the necessary (but ultimately liberating) deadlines to keep the chapters flowing. Tim Lootens and Karen Oswald of Lucent Technologies suggested CTI solutions for real-world problems, including a host of new applications. Many thanks to my staff at Enron Corp., including Juanita Mendez and Justo Morales. Other contributors include Kenny Rogers of AT&T, Frank Marino of Southwestern Bell, Coralee Jones, and Charles Reyes. Finally, the support of my family during the months of my "virtual absence" proved invaluable.

William A. Yarberry, Jr.
Houston, Texas

Chapter 1
Telephony Basics

CTI (computer telephony integration) applications must mesh with the existing legacy telephony environment. Organizations typically amortize their PBX investment between five and ten years, so new applications will need to fit the older telephony standards for the years to come.

The sections below outline the fundamental telephony components upon which CTI protocols and applications reside. Although IP telephony will eventually change much of the technology within the PBX and peripherals, many of the concepts will not change — because the same logical functions must still be accomplished.

A BRIEF HISTORY

Alexander Graham Bell, a Scottish-born American inventor, patented the first commercial telephone on Valentine's Day, 1876, just two hours before a similar patent was filed by Elisha Gray of Chicago. By 1878, the first commercial telephone exchange was brought into service in New Haven, CT.

The earliest telephones were sold in pairs — the purchaser was supposed to run wires directly between the two locations that needed to communicate. It did not take long for the nascent telephone company to realize that private wires strung over trees and buildings were going to be impractical if deployed on a large scale. In fact, to have all points on a network talk to all other points, $N * (N - 1)/2$ connections are required, where N = the number of points. Hence, the development of the "switch" that allows circuits to be opened and closed from a central point rather than having "nailed up" point-to-point connections everywhere.

The first CO (central office) switches were operated by young men (who soon proved too rude and were replaced by young ladies) connecting circuits physically with wires. Later, the automated switch allowed dialing without operator intervention. Eventually, the technology found its way into private businesses with large offices and the PBX (private branch exchange) was established.

The PBX, at its most basic, offers line consolidation and reduction in resources required from the CO. Generally, one can assume a 1:10 ratio of employees who are using the telephone (off hook) versus those who are not.

Thus, if an office building has 2000 employees, only 200 circuits from the CO are needed at one time. There are circumstances that require "non-blocking" circuits (e.g., emergency lines, the CEO telephone, key employee's telephones). Of course, this model breaks down under unusual circumstances, such as a disaster (all employees trying to call relatives and friends at the same time).

U.S. CARRIER STRUCTURE

There are two classes of carriers in the United States: local exchange carriers (LECs) and interexchange carriers (IXCs or IECs). Southwestern Bell, Bell Atlantic, and Powell Telephone Company in Tennessee are examples of LECs. IXCs include AT&T, MCI, and Frontier Communications. When AT&T divested its 22 Bell System operating companies, they were eventually regrouped into seven regional Bell Operating Companies (RBOCs). The RBOCs offer intraLATA services — long distance service within a limited geographic area.

Since the 1996 Telecommunications Act, the term CLEC (for competitive local exchange carrier) has entered the mainstream. CLECs either build their own infrastructure for local service or lease local loops from the existing LEC. Various cable TV, electrical power, ISPs, and cellular telephone companies have established CLEC subsidiaries.

Most carriers now use digital switches in their central offices. Some of the most common switches include:

- Lucent 5ESS
- Northern Telecom DMS-100 (the SL100 is the commercial premises version of the DMS-100)
- Siemens EWSD
- NEC NEAX 61E

NORTH AMERICAN NUMBERING PLAN

Dialing plans are critical — both for the public network and individual organizations. With the proliferation of cellular telephones, second lines, pagers, along with normal business growth, the North American Numbering Plan (NANP) has received considerable attention in recent years. Each time a new area code is added, routing tables of the PBX must be updated.

NANP is structured as follows. A ten-digit dial plan divided into two parts. The first three digits are the Numbering Plan Area (NPA), more commonly known as the area code. The remaining seven digits are also divided into two sections. The first three numbers denote the central office (CO) code, and the remaining four digits represent a station number (extension).

NPA (area codes):

- N is a value of 2 through 9.
- The second digit is a value of 0 through 8.
- The third digit is a value of 0 through 9.

A "1" in both second and third digits has a special significance:

Number	Meaning
211	Reserved for future use
311	Reserved for future use
411	Directory help
511	Reserved for future use
611	Repairs
711	Reserved for future use
811	Business office of CO
911	Emergency

There are also service access codes in the 700, 800, and 900 series (for toll-free and surcharge services).

NANP also provides additional central office codes:

Number	Meaning
555	Toll directory help
844	Time
936	Weather
950	Access to interexchange carriers (IXCs) under Feature Group "B" access
958	Plant Testing
959	Plant testing
976	Information delivery services

There are some special two-digit prefix codes used (other than the usual "1" for long-distance services):

Number	Meaning
00	IXC operator help
01	International direct-distance dialing
10	Used to dial "equal access" IXC. In most areas of the United States. callers can pick their IXC of choice by dialing an access code in the form 10XXX. See below.
11	Custom calling service

To allow callers to choose their long-distance company (e.g., use MCI from a telephone that would normally use AT&T if the caller dials "1"), the following equal access codes have been defined (new ones are being added routinely):

Number	Meaning
031	ALC/Allnet
222	MCI
223	Cable and Wireless
234	ACC Long Distance
288	AT&T
333	Sprint
432	Litel
464	Wiltel

INTERNATIONAL DIALING PLAN

The ITU-T (International Telecommunications Union-Telecommunications Services Sector) based in Geneva, Switzerland, established a dialing plan in the 1960s:

Zone (1st Digit)	Meaning
1	North America
2	Africa
3	Europe
4	Europe
5	Central and South America
6	South Pacific
7	Russia
8	Far East
9	Middle East and Southeast Asia

After the first digit noted above, each country will have a one- to three-digit country code assigned to it. Each country has an access code that callers must use to dial international calls (some countries have the same access code; the United States and Canada, for example, both use 011 for international calling).

All the above dialing plans address the *public* network. Many larger organizations with dedicated links (or merely IP connection points over the Internet) will have a private dialing plan that minimizes costs (least cost routing) and simplifies dialing for the end user.

THE TELEPHONY PROCESS AND SIGNALING

The fundamental processes of calling were established in the late 19th century; in some ways, changes have been slow in the ensuing 100 years. Following is a synopsis of the basic call flow for telephony:

- Answering a call. When a call comes in on an analog telephone, the PBX at the central office applies a ring voltage of approximately 70 to 90 volts to the telephone circuit. The telephone rings. When the recipient of the call takes the handset off hook, the CO determines that the circuit is complete and the phone receives an analog signal. For digital telephones/transmissions, the computer telephony (CT) application recognizes the incoming call, takes the telephone off hook, and receives digital data.
- Identifying the caller. Although many older telephone networks do not support Caller ID or DNIS, this capability is rapidly becoming more common in the telecommunications world. If Caller ID is supported, a CTI application (e.g., Phoneline from CCOM) can pick up the name and number of the caller during ringing. This number can be used for accessing databases and doing screen pops so that the called party knows in advance who is calling and perhaps some historical information. There are four potential ways to identify a caller: (1) Caller ID (analog); (2) automatic number identification (digital); (3) direct inward dial — an organization may have a specific DID assigned to a particular set of callers (customers, suppliers, etc.); and (4) dialed number identification service (DNIS). Note that DNIS allows a single trunk group to be shared by multiple 800 numbers and still allow the recipient of the call to identify the number originally dialed. An order fulfillment organization, for example, may have one toll-free number, 888-123-1234, for gardening supplies and another number, 800-345-3456, for painting services. Both numbers could be terminated at the same agent's telephone and DNIS would let the agent know which product is of interest to the caller.
- Making a call (analog). The process is to initiate the call and then dial. Call progression tones reflect dial tone, ringing, busy, fast busy, etc. Analog systems use two common methods to initiate a call: (1) loop start and (2) ground start. A phone line is seized (started) by giving it a supervisory signal. Loop start inserts a resistor between the two wires ("tip" and "ring") of the telephone when the receiver is lifted. When the central office detects loop current, a dial tone is generated. Loop start is typically used in homes. The ring lead is connected to −48 volts and the tip lead to connected to ground. When the central office detects the loop and the fact that it is drawing dc current, the ringing at the other end stops. A ground start architecture functions by grounding one of the two telephone wires when the receiver is lifted. When the central office detects a ground wire, a dial tone is generated,

enabling outward dialing. One last type of signaling is E&M (historically, "ear = receive, and mouth = transmit"), which is used often for two-way switch-to-switch or switch-to-network connections.

- Making a call (digital). Wink start is the most common method for digital telephone systems to start a call. Although the end user hears a dial tone when the handset is picked up, the PBX operating system actually sets a bit to signal the central office that the line is now active. The central office responds with a short-duration off-hook pulse or wink (usually 140 milliseconds) that tells the phone or telephony board inside a PC that it is ready to receive a dialed number for out-calling.
- Dialing. The two common methods for dialing are touch tone or DTMF (dual-tone multiple frequency), and rotary (still used by 60 percent of users worldwide). The keys on the telephone instrument have two standard frequencies per key. For example, when "1" is pressed, frequencies of 1209 Hz and 697 Hz are sounded simultaneously (see Exhibit 1-1). DTMF is clearly more efficient than rotary and can be transmitted over virtually any media (microwave, copper twisted pair, and carrier facilities). In addition, it is resistant to interference from line noise. Voice mail applications, such as Siemens' Phonemail, use touch tone digits to communicate with the handset (for saving, erasing, and giving instructions).

As the number is dialed in a rotary instrument, the circular dial sends out a momentary break in the DC circuit equal to the digit (three opens and closes for the number 3). The CO detects the evenly spaced opens and closes and registers the appropriate digit. Rotary dialing is considerably

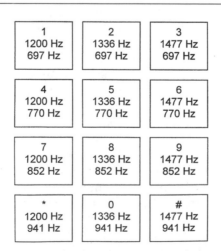

1 1200 Hz 697 Hz	2 1336 Hz 697 Hz	3 1477 Hz 697 Hz
4 1200 Hz 770 Hz	5 1336 Hz 770 Hz	6 1477 Hz 770 Hz
7 1200 Hz 852 Hz	8 1336 Hz 852 Hz	9 1477 Hz 852 Hz
* 1200 Hz 941 Hz	0 1336 Hz 941 Hz	# 1477 Hz 941 Hz

Exhibit 1-1. DTMF frequency groups.

slower than DTMF dialing, and some COs do not recognize the protocol. A robust CTI application will have the capacity to recognize rotary digits and voice-initiated digits in order to get around these limitations.

Another dialing protocol is multifrequency (MF) dialing, used for internal signaling at telephone companies. Also called multifrequency pulsing, MF uses the standard ten-decimal touch-tone digits plus five auxiliary signals to generate a specific sound. MF signals received considerable publicity during the 1960s as a result of the "Captain Crunch" toll bypass scheme. The following synopsis is from Harry Newton's *Telecom Dictionary*, 14th edition:

> At one point in the 1960s, a breakfast cereal had a promotion. It was a toy boson's whistle. When one blew the whistle, it let out a nearly precise 2600 Hz tone. If one blew that whistle into the mouthpiece of a telephone after dialing any long-distance number, it terminated the call as far as the AT&T long-distance phone system knew, while still allowing the long-distance connection to the distant city to remain open. If one dialed an 800 number, blew the whistle, and then pressed in a series of tones (called multifrequency or MF tones) on your "Blue Box," you could make long-distance and international calls for free because the only thing the local billing machine knew about was the original toll-free call to the 800 number. ... Cap'n Crunch's legacy (he was put in jail four times during the 1970s) is System Signaling 7, a system of immense benefit to all.

- Call progression. The public network (PTSN) and most PBXs use audible tones to indicate the movement of a call toward completion. For example, tones of dialing, busy, ring-back, error tone, and ringing are common tones used to tell the user (and internal systems) what is occurring. Computer telephony boards use call progress tones to determine the state of a particular call. A predictive dialing application, for example, will need to determine the state of an outbound call (if a busy signal is received, go on to the next number and do not send to the agent). A call progression tone varies in frequency and cadence. It usually has two frequencies (ranging from 315 Hz to 650 Hz) and the cadence is an alternating pattern of on and off. Desirable features of call monitoring equipment include the ability to detect stutter dial tone and elimination of noise (via filters or software algorithms). Note that tone types will vary according to country and PBX. Any computer telephony boards used must be trained for the specific PBX to which they are attached.
- Termination of a call. On an analog telephone, when the receiver is put down, the loop current is terminated and the CO responds by terminating the line. Digital systems recognize an on or off hook bit and stop sending current to the line when the receiver is placed back on the base station.

Signaling between switching offices can be provided on a per-trunk or common channel basis. In common channel signaling (CCS), all the signaling information for one or more trunk groups is carried over a separate (not traffic bearing) channel. In a T1, for example, one of the channels would be for signaling and the remaining 23 would be available to carry traffic. Nonassociated CCS uses completely separate trunks or channels to provide signaling information. It is more economical and more flexible. Assume, for example, that a business has a promotion that causes lines to be swamped. With traditional CCS, available circuits may be swamped with call attempts, thus reducing the total capacity of the trunking available to an organization. Nonassociated CCS, on the other hand, ensures that load-bearing trunks do not get swamped by call attempts.

COMMON CIRCUIT CONNECTIONS: PBX TO CO

There are an increasing number of ways that the CO can be connected to the organization's premises equipment (via a demarc). Some of the more traditional methods include:

- Ground start, two-wire CO trunks. They can be incoming, outgoing, or both.
- DID CO trunks. DID means direct inward dialing and allows callers to directly reach extensions within the organization's workplace without the intervention of a human or automated attendant. DIDs can be ground start but more often are digital DS1 circuits for larger organizations. Another alternative is E&M signaling with four-wire circuits (now less common).
- WATS (wide area telephone service). This term is not used as much now as in the past. AT&T, to their later chagrin, forgot to trademark this name, so it is now in the public domain, meaning discounted toll service provided by long-distance and local exchange companies. Incoming and outgoing services were separate in the past but now can be commingled in the same trunk group (e.g., Southwestern Bell offers "smart trunks" that allow incoming/outgoing on the same trunks, increasing throughput by roughly ten percent). In the past, WATS lines were billed at a flat rate and many employees thought the calls were "free." That is no longer the case — all are on a per-minute basis. WATS trunks do not have addressable PSTN numbers and hence cannot be dialed directly from the outside. Hence, they are often implemented as hunt groups within the PBX. This sometimes causes a problem with caller-ID equipped users who receive calls at their home from small businesses or friends. When they receive a call from a business that has WATS lines and see a trunk group ID, they believe they can just return the call by dialing the number — of course, they get an error tone because the trunk groups are not DIDs and cannot be dialed directly.

- IXC trunks. Many businesses will have dedicated trunks to their long-distance carrier. These trunks may physically come from the LEC, but logically they are a pass-through to the IXC point of presence. With these arrangements, long-distance calls out and in bypass the LEC (from a billing perspective) and thus significantly lower long-distance costs. When an organization with a dedicated IXC trunk talks to another similarly equipped organization, the call is "dedicated to dedicated" and results in a low per-minute charge.
- FX (foreign exchange) circuits. FX circuits typically use a two-wire loop start configuration that originates from an LEC outside the organization's subscriber exchange area. Sometimes, this is done to avoid long-distance charges and sometimes as a disaster recovery option. Assume, for example, that a Houston-based hospital is serviced by a Southwestern Bell CO located on Jefferson Street. The hospital has a vital need to stay connected and cannot afford to be incommunicado if the Jefferson Street CO has a fire, terrorist damage, etc. One method (and there are several) to ameliorate the risk would be to run FX lines from another Southwestern Bell CO in Houston (e.g., the Galleria CO). FX can refer to a few lines or a series of trunks. If trunks are involved, then the connectivity is not simply from a foreign CO to a telephone set, but provides connectivity to a switch.
- Tie lines. Still heavily used, tie lines are direct, dedicated circuits from one location to another. They can be ISDN digital circuits, analog lines, microwave, or other types of links. Alternatives to tie lines include meshed networks (such as Frame Relay) and the Internet (IP network).
- OPX lines. Off-premise extensions (OPX) are telephones that are not physically close to the PBX that provides their service. Generally, the PBX requires an OPX card and there are distance limitations (in terms of electrical resistance).
- Automatic ring-down circuits. Ring-down circuits are used in situations where a user needs to immediately call a specific location without the need to dial a number. For example, when someone is stuck in a defective elevator, he or she picks up the elevator telephone and immediately the telephone on the service desk rings. Traders connected to the New York Stock Exchange may have similar arrangements. In a ring-down circuit, an ac current is sent down the line (local or long distance). The current may light a lamp or ring a buzzer. Ring-downs are expensive but appropriate for many business/operational environments.
- Wireless options. In some situations (e.g., rural environments), landlines may not be the easiest way for a PBX to be linked to the LEC. There are a number of wireless options, including satellite communications (VSAT), microwave, or line of sight infrared. Except for VSAT, the user should not perceive any delay in ordinary conversation.

COMMUNICATIONS CONCEPTS AND TECHNIQUES USED IN TELEPHONY

As communications technologies have matured, layer upon layer of abstraction has been added to the sum total of communications infrastructure. For example, a perusal of Claude Shannon's 1948 paper on information theory is a daunting task for all but the most accomplished mathematician. Fortunately, as technologies have evolved, so have standards and intelligent software. For the communications manager and project leader, it is often enough to understand the concept and the available products. Following are some of the communications techniques used in telephony systems and networks.

- CODEC. A CODEC (COder-DECoder) converts a signal from analog form to digital signals that can be used by modern PBXs and transmission devices (e.g., Cisco 3810 ATM boxes). The same equipment/algorithms convert the signal back in the speaker or earpiece so that humans can understand the message. In some cases, the CODEC is located in the PBX; and in other systems, the handset contains the hardware necessary to do the conversion. The term CODEC has most recently been associated with videoconferencing, where it refers to compression and conversion to digital form for transmission over long-distance lines. Exhibit 1-2 illustrates basic CODEC functions.

- Pulse Code Modulation (PCM). PCM is the most common means to encode a voice signal into a digital bit stream. For toll-quality speech, an analog signal is sampled 8000 times per second using eight bits to record the results. Only 4 kHz of bandwidth is needed to produce good-quality speech. Note, however, that since there are only 256 possible combinations of an eight-bit binary number, speech that is digitized can never be "perfect" using this scheme. Of course, it is certainly adequate for most purposes. Speech is understandable (but not pleasant) at sampling rates/compressions much less than 8000 per second. There are two implementations of PCM: m-law PCM and A-law PCM. m-law is used in North America and Japan and uses a process called companding to enhance the signal-to-noise ratio (companding compresses the amplitude range for economical transmission and then expands it back at the receiving end). In Europe, a slightly different companding technique is used, resulting in a different standard called A-law. Both provide for excellent voice quality and modem transmissions.

- Voice coding/compression techniques. From the viewpoint of the organization trying to minimize communication costs, reducing the bandwidth from 64 Kbps to a lower rate can be attractive. Exhibit 1-3 lists some common techniques for compression and bandwidth conservation. This is an area of intense research due to speech storage and bandwidth abatement.

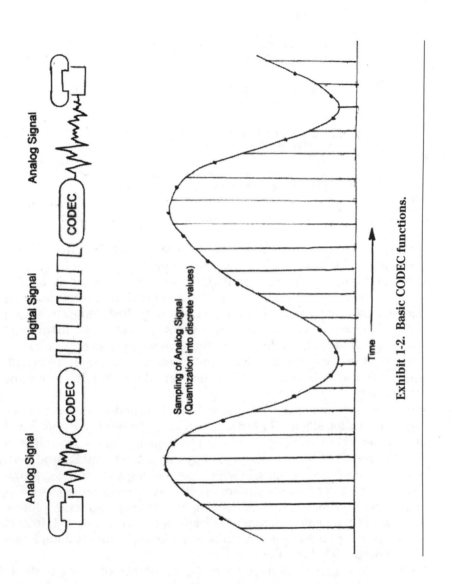

Exhibit 1-2. Basic CODEC functions.

11

COMPUTER TELEPHONY INTEGRATION

Exhibit 1-3. Techniques for voice coding and storage.

Technique	Method	Quality
PCM (pulse code modulation)	8000 samples per second. 64 Kbps 64 Kbps	High
ADPCM (adaptive differential pulse code modulation)	Compresses to approximately 32 Kbps. Calculates the difference between two consecutive speech samples	High
LPC (linear predictive coding)	Digitizing technique that drops voice to 2.4 – 4.8 Kbps.	Low
VSELP (self-excited linear prediction)	Digitizing technique used in digital cellular telephones. Compresses to 4.8 Kbps	Low
CLEP (code-excited linear predictor)	Government standard for compression. Compresses to 4.8 Kbps.	

- Circuit concepts. A channel represents a path of a single logical communication. Sometimes, a channel is a circuit, but a circuit is more often considered a physical configuration of equipment. A T1, for example, has 24 channels on one physical medium. Circuits can be simplex, carrying a signal in one direction only; half-duplex, carrying a signal in two directions, but not at the same time; and full-duplex, carrying signals in both directions simultaneously. Full-duplex speakers in conference rooms are significantly better, for example, than half-duplex speakers because of the tendency of speakers to interrupt each other.
- Digital versus analog signals. Circuits are designed to support either digital or analog signals. In order for an analog circuit (e.g., one used at a residence with analog telephone) to transmit data, start and stop bits are required to keep the signal synchronized (see Exhibit 1-4). Synchronous transmissions, in which the sending and receiving terminals receive a continuous stream of bits, are considerably more efficient (used, for example, in ISDN). Digital circuits also have a lower error rate than analog circuits. Typically, a bit error rate of 1/1000 is reasonable for analog voice circuits; such an error rate is completely unacceptable for data circuits.
- Multiplexing. Digital circuits have the capability of using a single broadband signal to carry several channels over a single circuit. A multiplexor combines the channels on the sending end; and at the receiving end, the signal is demultiplexed to restore the original channels. The composite signal contains data from all the end users. It provides for efficient use of transmission capacity. Although ATM is

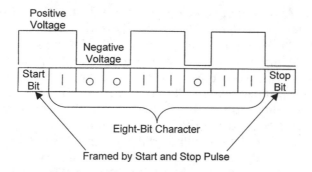

Exhibit 1-4. Asynchronous transmission.

increasingly used for wide area connections because of its ability to combine voice, data, and video, multiplexing is still widely deployed and is appropriate in many situations (ATM, for example, has at least a 10 percent overhead). The most common multiplexing scheme is TDM (time division multiplexing). TDM allows for a variation in the number of signals being sent along the line and constantly adjusts the time intervals to make optimum use of available bandwidth. It is also protocol insensitive, and can combine a number of different protocols within the same high-speed transmission link. Another scheme, FDM (frequency vision multiplexing) is used for cable TV where different stations are assigned frequency bands on a single cable medium.

- Basic carrier systems. Exhibit 1-5 shows the digital signal hierarchy most commonly used in North America. PBXs are linked to the LEC or IXC via channel service units (CSUs) and possibly data service units (DSUs). More recently, access concentrators are being used to link various voice over data network circuits (ATM and Frame Relay). During installation of new premises equipment, one of the most critical tasks is ensuring that all the protocols and parameters are set correctly between the central office and the premises equipment. Carriers

Exhibit 1-5. North American digital signal hierarchy.

Designation	Bandwidth	Number of DS0 Channels
DS0	64Kbps	1
DS1	1.544Mbps	24
DS3	44.736Mbps	672
OC12	622.08Mbps	9,344
OC48	2.4Gbps	37,378
OC192	9.6Gbps	149,512

(notably AT&T) have developed various T1 framing formats. Super-frames have been defined to include 12 frames at a time; extended superframe formats vary in length from 12 to 24 frames. Various schemes have been implemented to ensure the integrity of transmissions over long distances or with lengthy strings of zeros (found in data rather than voice transmissions). Some of these schemes include bipolar eight zero substitution (B8ZS) and robbed bit signaling (RBS). Premises equipment, using CRC (cyclic redundancy check) algorithms, can detect degradation in lines prior to complete failure.

- ISDN. Integrated Services Digital Network is worthy of a book in itself. The ITU–TSS (previously the CCITT) defines ISDN as a network service that provides end-to-end digital services (both voice and data) to end users. The original objective of the designers of the service was to rid the world of the inefficient and costly analog infrastructure that has been built over the years. ISDN does not require a massive investment in infrastructure as perhaps fiber to the residence would incur. It is deployed in Europe and in the more urban areas of the United States. Some of the benefits include: (1) the number and possibly the name of the calling party is transmitted before the call is answered; (2) by using two "B" channels, both voice and data applications can be run via the same circuit; (3) digital services are on a line-by-line basis; (4) dial-up calls are much faster than with analog (POTS) lines; and (5) additional information can be transmitted along the signaling D channel.

There are two implementations of ISDN. The lower bandwidth form, typically offered to residential customers of LECs, is called BRI (for basic rate interface). It has two B or bearer channels of 64Kbps capacity each, plus a D signaling channel of 16Kbps. Exhibit 1-6 shows a simplified BRI configuration. PRI (primary rate interface) carries 23 B channels plus a D channel (the international standard for PRI is 30 B channels and one D channel).

PHYSICAL COMPONENTS OF THE TELEPHONY SYSTEM

There are thousands of specialized components of a large PBX. Following are the most important.

- Digital stations. Virtually all major vendors offer digital stations to the end user. By converting the analog signal to binary format (via signal sampling at approximately 8000 times per second), each handset can perform a myriad of complex telephony functions. Unfortunately, many of these functions are unknown to the average user — CTI provides a means to make these features visible and far easier to use. Digital stations offer more complex features than the older analog stations (which do not have the advantage of a separate signaling channel and two digital bearer channels).

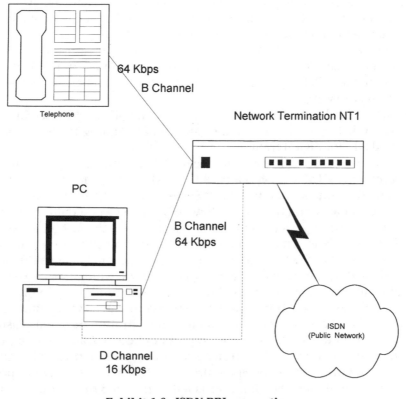

64 Kbps
B Channel

Telephone

Network Termination NT1

PC

B Channel
64 Kbps

ISDN
(Public Network)

D Channel
16 Kbps

Exhibit 1-6. ISDN BRI connection.

- Main cabinet. The cabinet houses the CPU(s) for system operation supervision, disk drives for the configuration and saving system transactions, memory, and bus slots for all the cards. Many systems are modular so that additional cabinets can be added as the need arises. One of the selection criteria for a PBX is whether additional stations (or analog lines) can be added without a "forklift" upgrade.
- T1 cards. Interface cards (DS1 cards). T1 lines travel from the demarc (of the local exchange carrier) to these interface cards. They are usually connected to the switch by an Amphenol connector. Twenty-four channels are available for voice, data, or video communications. These channels can be configured by the technician for any combination of traffic (of course, the CO must make the appropriate changes in its switch to match the requirements of the organization). For example, 12 channels could be allocated to AT&T for long distance, eight for Southwestern Bell local service, and the remaining four channels could serve as a tie line to another PBX in the organization.

- Input/output cards. These cards allow easy access to the switch for call detail (CDR port) and general I/O. Many switches now have an Ethernet connection (obviously important for any CTI application). Older switches, such as the Siemens Rolm 9751, have an SMIO (system management I/O) port that transmits screen dumps in response to line commands. Getting these screen dumps in a database format is not straightforward.
- Analog cards. If the organization uses analog trunks, these cards are needed. They serve the same function for analog traffic as the T1 cards do for digital traffic.
- Station line cards. These are the cards that connect to the wiring that goes to the frame. Typically, 8 to 24 stations (telephones) can be served by one station line card. Some switches have stations with an analog adaptor on the back, allowing both voice and data to be transmitted on a single telephone pair — back to a single port on the station line card. Thus, a user could have a PC modem connection and a digital set serviced by the same port. This is possible because the protocol used by most vendors is a variant of ISDN, which has two 64Kbps B channels; one channel can be used for voice and the other for data.
- System administration terminal. System commands, via command line or GUI, are entered into the system administration terminals (usually PCs with modems). Formatted reports of hourly traffic data on engineered resources such as trunk groups are generated via this link.
- UPS (uninterruptible power source). Switches are sensitive to power spikes, lightening, etc. If power is fed from a UPS rather than straight from the local power grid, electrical failure is less likely.
- Rectifier. Most switches operate on dc power. Accordingly, incoming ac power must be converted to dc via a rectifier.

CIRCUIT SWITCHING

Switching establishes a temporary or permanent connection between lines and trunks. A switch matrix generally uses a multiplexing technique known as TDM (time division multiplexing) to carry signals and messages from one point to another. In TDM, the information is carried in time slots arranged like soldiers marching single file across a bridge. The matrix ensures that each input signal will match the correct output signal.

All switches are driven by software. Traditional PBXs use a variety of (highly proprietary) UNIX operating systems, whereas some of the newer (and smaller) switches run on NT. As switches assume more complex operations, the likelihood of failure from software errors will undoubtedly increase.

"Line side" is the term used to indicate a connection between the PBX and the customer side. Typically, a line side port will link to either an analog (POTS) station or a vendor-specific digital station. The opposite of line side is CO side, which refers to links (trunks) that connect to a central office. Digital trunks are considerably more efficient than analog and offer superior error correction, signaling, and monitoring.

Other basic functions of a circuit-switched PBX include:

- station-to-station dialing (typically using a four- or five-digit internal dialing plan)
- direct inward dialing (DID), which allows stations to ring without operator intervention
- direct outward dialing (DOD)
- call detail, an audit trail of calls made to and from a station
- automatic call distribution (ACD); calls to a single number are distributed to a number of specified extensions
- voice mail
- directory services
- internal reporting and administration

SERVICE/FEATURE COMPONENTS OF THE TELEPHONY SYSTEM

Feature set is the arena of fierce competition between vendors. The major vendors have hundreds of features (of which most users know only five to ten) that they use to differentiate their products. Exhibit 1-7 lists features available from most vendors. Note that this list is only a small fraction of those typically available (see Appendix A at the end of the book for a more complete listing).

SUMMARY

This chapter addressed the basics of traditional telephony. Although the specific technology is quickly changing, much of the terminology and many of the functions will simply assume a new form in the IP/object-oriented/CTI/software-driven world of the 21st century.

Exhibit 1-7. Features available from most vendors.

Feature	Comments
Automatic call distribution (ACD)	ACD groups are the *sine qua non* of help desks and call centers. The PBX (in some cases, an adjunct server) receives the call and makes an intelligent decision where to place that call. ACD functionality allows the PBX to track calls by agent, gives the caller the option to leave messages, and distributes calls based on criteria such as area code. For example, calls coming in with area codes east of the Mississippi River could be sent to the "East Desk" sales representatives, and area codes west of the Mississippi could be sent to the "West Desk."
Automatic number identification (ANI)	Also referred to as "Caller ID" for analog circuits. With this feature, the PBX can pick up the incoming digits and perform some function with them. For example, the incoming number can be displayed or can be used to do a database lookup on a separate database server. ANI can be provided to the PBX by any CO switch that is SS7 (signaling system 7) compliant and does not require ISDN.
Automatic wake-up	Hotel industry feature offered by most major PBX vendors.
Call blocking	Allows PBX to block specific country codes or area codes (e.g., 900 calls).
Call coverage	The specification in PBX software of what actions to be taken when the called party does not answer his or her phone. Does the call go to voice mail? Is it sent to a location on the other side of the world (for around the clock coverage)? Does it ring at co-workers' desks?
Call forwarding	Also referred to as "send all calls." The user can enter codes to send any incoming call immediately to a designated number or voice mail or some other receiving function (such as an IVR). This saves callers time. If users know they will be out of the office, they can set the appropriate codes and callers will not be forced to sit through four rings to get voice mail.
Call pickup	Allows any user to answer someone else's ringing telephone, so long as both parties are in the same "pick group." Typically used by an administrative assistant who is covering for a number of other users.
Camp on	Ability to dial an extension and if that extension is busy (in use), the PBX automatically continues to try to reach the extension, although the originating caller has hung up. Once the dialed extension is available, the PBX rings the originating party and the called party. This saves the user the effort of continually trying to reach a busy extension.
Class of service	A code assigned to each extension that specifies the PBX functions allowed. For example, lobby telephones may be assigned class of service 5, which allows local calls only. Each organization can define class of service to meet its needs. It is a good security practice to assign only as many features to a class of service as are needed for the job.
Conference calling	Allows the user to initiate a conference call for a maximum number of internal and external callers, depending on vendor. The sound quality will typically not equal that of a dedicated call bridge such as AT&T "meet me" service. In addition, typically less than a dozen participants can be accommodated from a user digital set. Conferencing on an extension does not have the volume and balancing controls found on a true "call bridge" device.

Exhibit 1-7 (Continued). Features available from most vendors.

Feature	Comments
Digit translation	The PBX should be able to translate an arbitrary number of digits (e.g., "12345") into a different number (e.g., "011-044-123-1234"). This capability includes masking — the ability to change all outgoing numbers starting with "6" to start with "2." Some systems lack the CPU and processing power to do a large number of digit translations without affecting systemwide performance.
Direct inward dialing (DID)	Parties outside the building can call the extension directly rather than having to go through the operator or an auto-attendant. The local exchange company usually assigns a range of DIDs that are available to the organization. Obtaining a contiguous range of numbers greatly simplifies day-to-day assignment of extension numbers to new employees.
Direct outward dialing (DOD)	Allows an extension or modem to directly connect with outside trunks. Modems are often put on DOD only lines; this increases security and avoids use of numbers in the DID range (which are sometimes scarce — particularly if an organization has grown rapidly at one site).
Directory	Allows user to scroll through names on a handset display in order to find the correct extension to dial.
Hunt group	An association of extensions such that when one of the dialed extensions is busy, it goes to the next, then to another, etc. Various PBX implementations determine the order of the search. For Help Desks, hunt group configuration should be carefully designed. For example, assume that agent #1, #2, and 3# receive all calls in a hunt group. If the PBX always goes to agent #1 first, then the call workload will be skewed — #1 will never get a break and #3 may be underutilized. Random starts would be preferable in that work environment.
	Hunt groups and ACDs share many similar attributes. However, hunt groups are more rigid in their function (e.g., always random or always go to agent #1 first). ACDs, on the other hand, can use more sophisticated CTI functions and capabilities within the switch itself to route a call.
Last number dialed	This should be available either as a standard button or as a programmable button on the handset.
Least cost routing	The PBX automatically selects the least cost route to send a phone call. The least cost could be tie lines, PSTN, Internet lines, a microwave circuit, or some other route. This feature is transparent to the user.
Message waiting light	This feature is obvious. It is the aesthetics that should be reviewed. Some users prefer large, highly visible lights; others like a tiny, discrete blinking light that can be more easily ignored.
Off premises station	For at-home or remote office workers, off-premise digital telephones are essential. The PBX has a card that is dedicated to off-premise extension telephones.
Paging	Implemented via telephone speakers using the paging channel. Usually a specific number is dialed and all parties connected can hear the message over a loudspeaker. There is also a common feature that sends a numeric page from voice mail based on preset criteria (e.g., page only when a high-priority message is received)

Exhibit 1-7 (Continued). Features available from most vendors.

Feature	Comments
Personalized ringing	Ability to modify the ring of the handset. Particularly useful in high-density cubicle work environments where several telephones may be ringing at once.
Recorded announcements	Announcements associated with voice mail should be flexible. Examples: customized greetings for internal and external callers, and the ability to use ANI (automatic number identification) to recognize the caller and play a special greeting (e.g., if an important client calls, the greeting could be customized to say "Ms. Jones, I'll be at 713-123-12345 if you need me right away"; all others would get the message "I will be out of the office until Monday. Please leave a message.")
Skills-based routing	Agents have different skills (e.g., speak different languages, have in-depth "tier two" experience, know a particular set of products). If those skills can be assigned to a database that can be queried, calls can be sent to the right agent. An IVR could say "Press 1 for an English speaking agent, press 2 for a Spanish speaking agent." The IVR/ACD software would use a database of skills to route the call to the appropriate agent.
Look-ahead routing	This PBX function checks for network busy conditions before attempting to complete the call.
Time-of-day routing	For organization's whose contract with their long-distance carrier has an off-peak discount, the system routes the call (to the extent possible) based on the time-of-day discount structure.
Speed dialing ("rep" dials)	Programmable rep dials are a convenience for the user. When reviewing the capabilities of a PBX, it is useful to ask whether user-programmed rep dials can be retained if the telephone is moved (either literally or virtually) to a new location.

Chapter 2
Voice Communications Architectural Design

CTI APPLICATIONS AS WELL AS GENERAL COMMUNICATIONS SERVICES ARE ENABLED by the architecture of the communications server and related adjuncts. When considering the purchase or expansion of a voice system, a multitude of characteristics must be considered in order to optimize performance and minimize costs. Following are key functions that can vary by vendor and should be reviewed in detail.

NETWORKING

Networking capabilities apply to both premises equipment and remote locations. An inadequate networking feature set limits growth, frustrates users who want similar functions in multiple offices, and increases expenses by forcing duplication of equipment. Important factors include:

- Protocol support. Is E1/T1 trunking supported? What versions of ISDN are supported (National ISDN-2, ISDN "call by call")?
- D channel sharing. Lucent refers to this capability as "nonfacility associated signaling." A medium to large communications server will have multiple T1s as DID or CO trunks. By using one D channel over multiple T1s, more bandwidth can be obtained for the same number of trunks. Exhibit 2-1 shows an example of the bandwidth efficiencies obtained using D channel sharing.
- N x DS0 support. By aggregating multiple 64K channels, certain applications (such as video) can be run at higher bandwidth. The session is treated as a single call (and may be billed at a lower rate by some carriers).
- ATM and Frame Relay Support. Some manufacturers provide cards in the communications server itself, whereas others rely on external boxes to provide the necessary connectivity (including conversion to and from TDM).

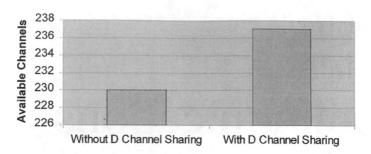

Exhibit 2-1. Effect of D channel sharing with 10 T1s.

- Maximum dialing string length. More is better. As the network gets more complex, more digits may be needed to traverse the devices on the other end; 28 digits or more is a reasonable benchmark.
- Universal dialing plan. Offices linked with dedicated lines can participate in a simplified dialing plan. For example, an organization with offices in London, New York, Qingdao, and Prague could establish a five-digit dialing plan so that employees can simplify their dialing. In addition, the dialing pattern is the same in each location. The communications servers must be linked and have the appropriate software in order to have this functionality.
- Communication server networking. In order to share central services such as voice mail, fax-on-demand, IVR, and other CTI functions, networking software must be in place to make the system of servers function as a single logical unit. Examples include NEC Fusion Call Control Signaling (FCCS) and Lucent Distributed Communications System (DCS). When linked as a single logical entity, users at one site dial users at other linked sites with the same dialing pattern. Also, voice mail functionality (transfer, mark private, return receipt, return call without requiring the number to be reentered, etc.) is transparent across nodes.

  Technical specifications for the links that support networking should be examined carefully. Some vendors support distributed communications links only via dedicated lines (digital or analog). Others may support links via dial-up ISDN over a carrier's virtual network (e.g., AT&T Software Defined Network). The advantage of virtual network support is that small offices in remote areas can be linked without the expense of a dedicated T1 access line. A campus environment may require a fiber-direct connection between servers. Not all vendors support direct connects via fiber.

- Central administration. Organizations with multiple office locations can obtain economies of scale by administering moves, adds, and changes remotely. In addition, system-level changes such as adding new North American Numbering Plan entries or making a dial plan change can be efficiently implemented via remote log-in. Products such as NEC AIMWorX or Lucent TERRANOVA provide the functionality to administer large networks of communication servers. Some features that should be reviewed include:

— Can a system-level change be promulgated across all servers with a single session, or does each site have to be individually administered?
— Does the software automatically add a voice mailbox to match the user's class of service?
— Are call accounting, trouble ticket, work order, cable management, inventory management, traffic analysis, and billing audit functions included in the base package?
— Are ports, lines, etc., shown graphically?
— Can telephones be initialized at a jack so that the extension rings at the appropriate physical location? This capability greatly reduces the need to run cross connects in the closet (called TTI for terminal translation initialization in Lucent documentation, and auto location reset in Nortel documentation).
— Can network monitoring products such as HP OpenView monitor the servers via SNMP?
— Are totals of lines, telephones (digital and analog), trunks, etc. shown?
— Are out-of-service trunks identified?
— Can information be imported/exported via standard PC-based files?
— Does the system exhibit referential integrity? For example, will it allow an extension to be deleted without sending an alert or at least an audit report showing hunt groups with extensions that no longer exist?

• QSIG. Standards bodies have developed a version of ISDN PRI that allows communications servers from different vendors to provide a collection of services that can be made available to users (see Exhibit 2-2). It is still a developing standard and does not provide the robust networking found in a proprietary link such as Lucent DCS. Nonetheless, QSIG provides significant benefits to corporate users in disparate systems:

— name identification
— call intrusion: a calling user can request immediate connection to a busy destination
— do not disturb: all incoming calls rejected by the QSIG network
— call waiting: users can accept, reject, or ignore an incoming call if they are already on a call
— call completion: when users dial a number that is busy, the request will automatically be completed when the line becomes free
— call forwarding when busy
— call forwarding if no reply
— call forwarding unconditional

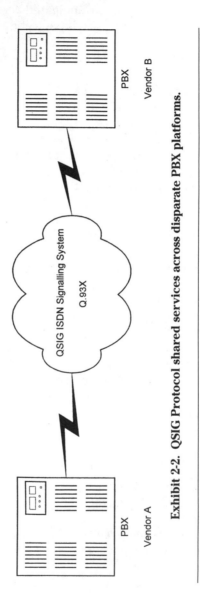

Exhibit 2-2. QSIG Protocol shared services across disparate PBX platforms.

— operator services: for example, a night attendant service could be shared among several sites

— other services are in development

PERFORMANCE FACTORS

Telephony hardware and software cannot be selected solely on the basis of features. Under conditions of high load, users may receive busy signals and functions may not be available unless the system has adequate bandwidth and processing power. Factors to be considered include:

- Processor speed. Proprietary RISC (reduced instruction set) chips have traditionally been used for faster processing, but off-the-shelf CISC (complex instruction set computing) processors from Intel are used in many switches. For example, Nortel uses off-the-shelf processors in its DMS-100 central office switch. Whether RISC or CISC, the processor can be a bottleneck if there are too many calls, ACD, or CTI events happening at one time.
- Busy-hour call completions versus attempts. Any benchmarking of call processing capabilities should include busy-hour completions rather than attempts. Also, the type of phone used in the benchmark should be considered. It takes considerably less horsepower for the CPU to communicate with an analog set than with a digital set.
- Processing beyond simple call connections. ACD, CTI, networking decisions, and other adjunct processing heavily exercise the PBX CPU and bus. Additional power is required to ensure smooth performance when these applications are running. Before evaluating alternatives, approximate busy-hour calls should be known, if possible. In addition, quantities of calls, duration of calls, number of agents, and number of call connections should be considered. In small offices with low activity, processor power is not significant.
- Non-blocking switch. Some vendors, such as NEC, claim that 50 percent of the users on one of their switches can pick up the telephone and dial the other 50 percent and actually get through. Trunks and lines should not contend for the same slots.

FAULT TOLERANCE AND REDUNDANCY

User expectations for telephony uptime are exceedingly high — 99.9997 percent reliability is merely adequate — particularly if critical functions are part of the business. PBXs, voice mail, and adjuncts vary in the number of redundant components and paths, as well as the speed with which they recover from a fault.

The key to maximum uptime is that all components, from the telephone set to the CO trunks to building power — should be equally robust. The old adage that "a chain is only as strong as its weakest link" applies to telephony as well.

Areas of critical importance for fault tolerance include the following:

- Duplicated processors and memory.
- Duplicated system bus and peripheral control processors. System bus redundancy allows processing to continue for all peripherals in the event of a hardware fault.
- Memory shadowing versus hot standby. Fault-tolerant systems, such as Nortel Meridian Supernode, use "shadow memory" to simultaneously process the same call in two locations in hardware. If one side fails, the call continues uninterrupted. Only call processing (the attempt to complete a call) must be reinitiated. Other systems will drop the current call while the switch to the other processor, memory, etc. takes place.
- Power supply. Power supplies should be duplicated. Ideally, duplicate power will be at the shelf level. Single points of failure should be limited to only individual T1 interface cards for trunks and line peripheral processors for stations.
- Surge protectors and UPS. Switches should receive only conditioned power, preferably through an uninterruptible power source (not merely public power with a battery backup).

GROWTH AND SCALABILITY

Most organizations dislike "forklift" upgrades. It means there is no way to gracefully expand the current system, resulting in a purchase of new hardware/software, disposal of old equipment, increased costs, and usually some disruption to users.

The first question to be asked is: how many boxes can one logically and seamlessly link together? Small offices, where key systems are often implemented, find themselves unable to expand beyond a certain number of cards (with limits on both digital stations and analog lines for modems and faxes). If small PBX systems (e.g., Lucent Prologix) can be linked, considerable growth potential is available.

The card/slot fit also affects growth. For example, NEC has a port interface module (PIM) with 18 slots. Each PIM can support a maximum of six 24/32 port digital trunk cards or digital remote units (DRUs). When the card limit is reached, another PIM must be purchased even if there are 16 time slot cards unused in an installed PIM. Ideally, slots should be "universal;" in other words, any card will work in any slot.

Software (CTI, ACD, etc.) should be portable to larger boxes within the vendor's product line. The converse is also true — applications developed at a headquarters location, for example, may need to be duplicated in smaller field offices.

Assuming that the organization wants to standardize on a single vendor or product line, scalability becomes important as communication systems are moved from one location to another. What is the effect of adding a large number of ports? Will the same systems work internationally as well as domestically?

Voice mail systems do not always scale well. As the number of users grow into the thousands, voice mail servers may need to be linked together. The Lucent Intuity Audix system, for example, uses an internal, intranode networking system called HICAP to present to the user an integrated, logically unified voice mail application. Some questions to be asked of the voice mail vendor include:

- What process is used to load balance the number of users placed on each server?
- Can users get into the voice mail system via a single access number?
- What effect do all employee broadcasts have on system performance? Most systems will experience a slowdown due to port contention and demands on the processors.
- Are there any difficulties or special functions needed to transfer an outside caller to a specific mailbox without requiring the caller to enter an extension number?
- Does the voice mail system have to be brought down for routine maintenance?
- Can the voice mail system survive a single server failure. In the case of Lucent Intuity Audix, for example, calls can be received even when one of the servers is down. However, employees must wait to retrieve their messages when the downed server is brought back into service.

TELECOMMUTING

Agents and employees need not be physically on premises to receive all the benefits of a digital communications system. Vendors offer "extenders" that use either an analog POTS line or an ISDN connection to homes or remote offices. These devices allow the user to use all the features available at the office. From the perspective of the PBX/telephony servers, the employee is on premises. A special card is required at the switch, and each telecommuter needs an extender at his or her residence.

In reviewing the architecture of the network, the treatment of off-net calls should also be examined. In many systems, once the call has been forwarded off-net, it cannot be returned if there is a "ring, no answer." Ideally, the last stop would be voice mail at the central site rather than ringing with no options for the caller.

Job sharing requires the same flexibility as telecommuting. If two individuals share a desk/telephone, they should be able to have their own voice mail, coverage path, greetings, and other customized features. By logging in, the system recognizes who is using the set currently and directs calls according to preestablished call processing parameters.

MANAGEMENT REPORTING/CALL CENTER FUNCTIONS

There is a surprising difference between vendors in ACD reporting and dynamic agent configuration capabilities. Refer to Chapter 9 on call centers for an in-depth discussion of management reporting and call center functions.

VOICE MAIL

When the PBX and voice mail are from the same vendor, the functions are usually tightly coupled. That is not necessarily a sufficient reason to implement like-vendor hardware. For example, an organization with a heterogeneous base of installed PBXs might want to install a standard voice mail (such as Centigram or Octel) for each site so that it would be easy to network the voice mail system. Otherwise, the only means of networking unlike voice mail systems is to use the AMIS (audio message interchange specification) protocol or VPIM (voice profile for Internet messaging).

Another important architectural decision is whether to centralize or decentralize voice mail services. Centralization provides for:

- economies of scale because the per-port cost goes down as volumes rise
- ease of maintenance, because the central location is most likely in a larger city, where maintenance services are more readily available
- ease of administration; distribution lists can be easily shared and security policies can be uniformly applied
- straightforward transfer of messages from one mailbox to another

The decentralized approach, on the other hand, has the following advantages.

- risk is distributed; if one voice mail server goes down, only a fraction of the user base is affected
- IXC channels are not tied up with constant voice mail traffic
- voice mail is not dependent on tie line availability
- local preferences, such as retention periods of messages before deletion, can be maintained
- hardware and software limits are not approached if the number of mailboxes is kept small

There is increasing interest in transferring voice mail over the Internet. Since the transfer of voice mail messages need not be isochronous, packet transmission is an ideal transport. Either IP over dedicated lines or the public Internet can serve this function, although the public Internet is not private.

Finally, when evaluating voice messaging, the following basic questions should be asked to ensure that the product meets minimum industry standards.

- Does the caller have to redial the called extension number after entering the voice mail system?
- Can the caller transfer out of voice mail?
- Do the PBX and voice mail systems communicate well enough to always provide a message waiting light (or stutter tone) when a new message has been received?
- Can a troubled port be put out of service so that calls are not placed to malfunctioning ports?
- Does the system have call disconnect supervision? In other words, will a port always be freed immediately when the call is terminated?

STATION FEATURES

Although the telecommunications group sees the internal features of the telephony system, users see only the physical station at their desks. Fair or not, many will judge the system simply by the appearance and functionality of the station. Key factors include:

- Number of buttons. In spite of alternatives, many users insist on using a digital telephone with a display like an old key system. They want to see every individual in their department as a separate key. Because the paying user is always right, this should be a consideration in selecting the station model. One option may be to purchase a standard model for the entire organization but ensure that add-on modules with extra buttons are available for those who need it.
- Speaker phone. Deluxe models (more expensive) have full-duplex, built-in speakers. However, even standard sets should have half-duplex speakers that can be easily interrupted. Some older systems require the user to almost shout to get the speaker phone to interrupt and switch the direction of conversation.
- Display information. The more information displayed, the better. ANI, directory name, status indicators, etc. are keenly appreciated by users. Display of help information saves many calls to the telecommunications Help Desk.
- Aesthetics. Sleek, modern, clearly labeled — all are important to user acceptance.
- Mundane items. Users care about cord length, color, shoulder rests, wall mount versus desk mount, and other seemingly trivial concerns. But it may not be as trivial as one would suppose. A commodities trader, for example, may have the following on his or her desk: a PC,

a Sun workstation, a turret trading telephone, a stenophone for direct communication with co-workers, an Amtel digital communicator, and a standard telephone. Desk space for this high income-producing individual is at a premium. Small is better.

VOICE OVER DATA NETWORKS

As the telecommunications industry moves toward packet transmissions for all traffic, it is clear that any communications system design today must incorporate voice over IP, ATM, and Frame Relay. Prudent questions to ask include:

- Is IP, ATM, or Frame Relay transmission based on a card or a separate server? If a separate server, how many ports are supported?
- What is the smallest configuration that will support these technologies? Does the vendor's smallest PBX support these technologies?
- What LAN connections are supported — Ethernet? Fast Ethernet? This is an important consideration because IP voice traffic that is generated by a "black box" Internet telephony server must necessarily reside on the LAN before it is sent to a router and then to the Internet (see Exhibit 2-3). If the LAN becomes saturated with traffic, there will be a marked deterioration in quality of sound at the receiving end.

PAGING SERVICES

In some organizations, such as a hospital, paging is a critical feature. Some points of differentiation between vendors include:

- Is paging solely attendant based or does it use a call park feature? Ideally, the attendant can answer an incoming call, park the call, and page the called party. The called party can go to the nearest phone, dial in a code, and be connected directly with the calling party without operator intervention.
- Can paging be segregated by floor or zone so that employees do not hear pages that do not apply to them?
- Does paging comply with the Americans with Disabilities ACT requirements, such as video text displays?
- Is priority paging supported?
- Can additional (third-party) paging services be added to the platform?

SECURITY FUNCTIONS

Although security will be discussed in depth in a later chapter (Chapter 11) the following should be part of an initial review of system capabilities:

- How are security alarms presented — via pager, terminal display, phone calls to the console, etc.? What events trigger those alarms? How are the controlling parameters set?

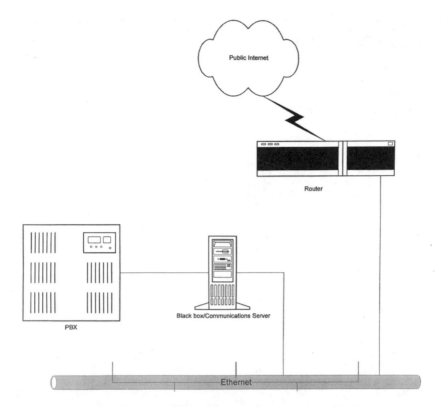

Exhibit 2-3. Telephony on the Internet using a communications server.

- What authorization codes are used?
- What password changes are forced (particularly maintenance ports)?
- Can internal and external (off-premise) forwarding be separated?
- How are 911 calls treated? Can specific room numbers be identified from the outside?

SUMMARY

The marketplace offers many permutations of cost, quality, and technical direction. By understanding the areas where hardware and software vendors differ, the telecommunications manager can better match the technology with the organization's needs. Some trade-offs are obvious — most key systems cannot be made to do ACD functions; highly redundant and fault-tolerant systems are only found at the upper end of the market; centralized voice mail requires long-distance tie lines or per-minute SDN costs. However, others require that the right (not so obvious) questions be asked.

Chapter 3
Interactive Voice Response Systems

INTERACTIVE VOICE RESPONSE (IVR) SYSTEMS ARE THE WORKHORSES OF THE TELEPHONE INDUSTRY. Everyone who uses a phone is familiar with the ubiquitous "For sales, press 1, for service press 2..." These systems provide answers to taxpayers preparing their tax returns at 2:00 a.m. on April 14th, send brochures via fax-on-demand to potential customers, use text-to-speech to provide the latest natural gas prices, and validate credit card numbers. Without IVR, the business (and non-profit) world would be slower, less efficient, and far more manpower intensive. Without IVR, many firms would be forced to close.

One reason IVR systems are so successful is that they are based on a simple input device — the telephone set. While the 12 buttons on the standard analog set may not be the ultimate in user friendliness, they are used constantly and are familiar. A well-designed IVR system requires only that the caller have a telephone and a POTS (plain old telephone) line. IVR applications with speech recognition capabilities even allow rotary phone customers to use the system. In short, IVR satisfies the three As — Anyone can use it, Anywhere, Anytime. CTI applications, on the other hand, usually require a PC to visually display information (although an IVR application can start up a CTI application).

The following chapter sections present a representative IVR system, some common IVR applications, development practices, hardware considerations, performance tips, and suggestions for improving the customer's IVR experience. This chapter includes only the highlights of IVR technology — it is rapidly changing (becoming less proprietary) and has a relatively low start-up cost, resulting in a plethora of vendors offering complete (albeit limited volume) systems for only a few thousand dollars.

A REPRESENTATIVE IVR SYSTEM

In order to provide a more real-world picture of IVR technology, the Lucent Technologies Conversant platform will be used as a model to illustrate hardware and software components that may be found in the market. Obviously, the IVR market is varied, in terms of call capacity, application features, database links, compliance with standards, redundancy, costs, etc. The Conversant is a high-end line that competes with Nortel Open IVR and similar products.

HARDWARE

Exhibit 3-1 shows the front view of Lucent Conversant hardware. The platform is linked to the PBX via a digital circuit card connected to a TDM bus cable or via a tip/ring circuit card over a standard telephone line. The Conversant can be used with a non-Lucent PBX by linking with the tip/ring circuit card (analog lines to the PBX).

Other components include a serial port, video port for a display terminal, keyboard port, floppy diskette drive, hard disks (connected via SCSI bus cables for higher speed), backup cartridge tape drive, and floppy diskette.

The base operating system is Unixware. Other IVR systems may operate on Microsoft NT or even Windows 98 (for low-end systems). The operating system must be able to efficiently multitask when many callers simultaneously access the system — particularly if they are multiple applications running. A later section in this chapter will address how the effects of memory, disk speed, and application design affect performance.

Much of the functionality in an IVR platform comes from the cards inserted into the chassis. The Conversant, for example, has a card for (not a complete list):

- circuits (both analog and digital, E1, T1)
- external alarms
- fax
- speech processing (signal processor)
- LAN connections (Ethernet or Token Ring)
- host (mainframe) interface
- PBX interface
- video

Applications (programs) are loaded onto the hard drive and run either partially or wholly in memory.

All the backup, redundancy, and performance concerns associated with any server apply to the IVR hardware as well, including:

- redundant power supplies
- backup CPUs
- fast backplane
- adequate memory (to prevent swapping/paging from the hard drive)
- mirrored hard drives (RAID compliance — data is written to the hard drive with error correction codes that allow it to be reconstructed if a single drive fails)

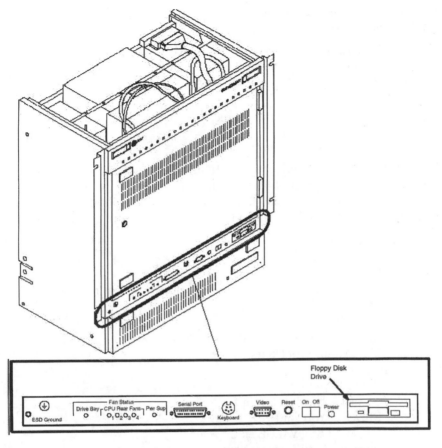

Exhibit 3-1. Lucent Conversant hardware platform (front view).

- multiple, independent cooling fans
- UPS, surge protector, etc.
- fail-over capability — the failure of one component does not bring down the entire IVR system
- self-monitoring capability, with alerts (low and high priority)
- hot swappable components (can replace cards on the fly)

DEVELOPMENT SOFTWARE

Example: Script Builder, Voice@Work and @Work Studio

Working with voice response applications is much like traditional client-server development packages. Actions by the user must be anticipated, data validated, database records updated, and appropriate response provided to the caller. The following excerpt from the Lucent Intuity Conversant System version 6.0 documentation demonstrates some of the techniques used to build an application.

> After giving a yes/no question, pause to give the caller time to respond, then present the possible answers. The prompt will stop playing as soon as the recognizer detects a spoken "yes" or "no" or a touch-tone signal. For example:
>
>> "You said 64587. Is this correct?"
>> [a 1.5 second pause]
>> "Please say yes or no."

Use the prompt and collect action to ask the question, play a series of silence phrases, then present the options. The figure below shows an example of how part of the Script Builder code will appear if one asks the caller for five digits, then confirms the entry within the prompt and collect action.

```
Prompt and Collect
    Prompt
        Speak with Interrupt
        Phrase: "please enter your 5-digit customer number"
    Input
        Mode: US_DIG
        Min Number of Digits: 05
        Max Number of Digits: 05
    Checklist
        Case: "Input OK"
            Speak with Interrupt
        Phrase: "You said"
            Field $CI_VALUE as C
        Phrase: "Is this correct?"
        Phrase: "sil.500"
        Phrase: "sil.500"
        Phrase: "sil.500"
        Phrase: "Please say yes or no."
            Confirm
        Case: "Initial Timeout"
            Reprompt
        Case: "Too Few Digits"
            Reprompt
        Case: "No More Tries"
            Quit
    End Prompt and Collect
```

Lucent's newest development packages, Voice@Work and @Work Studio, automate some of these tasks using graphical tools.

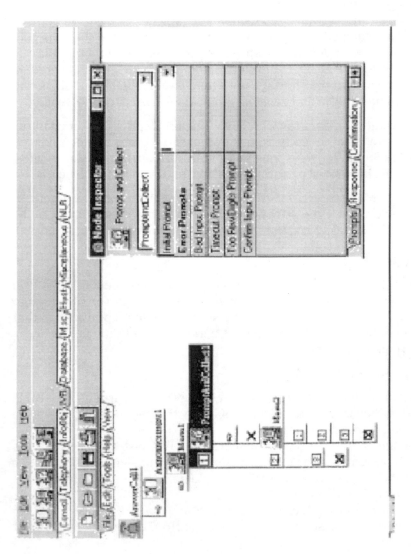

Exhibit 3-2. Sample Voice@Work development screen.

37

Voice@Work is an object-oriented development tool kit that uses Sun Microsystems Java and JavaBeans. With this toolkit (plus another package, @Work Studio), developers can write IVR applications without writing code. They can also port the same objects to several platforms, including IBM VisualAge. As will be discussed in later chapters, Java is platform independent and thus spares developers the need to rewrite applications for different vendor's hardware/software (see Exhibit 3-2). Voice@Work also includes a team development module that allows a team of developers to share speech, database tables, host definitions, speech recognition, and other resources. Another feature is the ability to develop remotely without impacting production. In practice, with the shortage of CTI programming/analysis skills, the ability to remotely log in and develop applications has become essential.

An important consideration (discussed in more depth in a later chapter) is the ability to generate code from the visual constructs (objects) developed. @Work Studio provides tools to generate call answer, business rules-based routing, integrated desktop, reporting, and data warehousing.

Given the proliferation of visual and object-oriented tools, the argument for in-house development of telephony applications is getting stronger all the time. And it is often faster and less expensive (after the initial learning curve).

APPLICATIONS

The above discussion outlined specifications of the Lucent IVR platform in order to give the reader a feel for the hardware and software required to implement IVR systems. The following sections reflect applications that are implemented by many IVR vendors across a broad range of business and government organizations.

The general benefits of IVR include:

- elimination or reduction of operator/agent time; in particular, it avoids peak staffing issues during times of heavy transactions (e.g., vacation season for travel agents)
- value added for wait time; a well-designed IVR system can tell callers they have X minutes waiting time for a live agent — would they like to hear about the company's newest products, answer a survey, or order products via touch tone?
- entertainment to reduce caller frustration
- ability to broadcast a repetitive message (to a voice mail system, for example)
- mechanism for casual users who, for example, would not want to dial into a Web site because they do not have a PC or it is too much trouble to boot up their PC for a simple transaction
- provision to supply constantly updated information via voice (especially with a text-to-speech facility)

- 24-hour-a-day, 365-days-a-year availability
- capture of customer/caller information that would otherwise be un-available
- streamlines operations

Specific Applications

The following are particular applications that apply across a number of business environments.

- banking by phone, including account balances, cleared checks, and fund transfers
- validation of credit cards and other financial instruments
- human resources: vacation and sick information, 401K info, savings plan, HR policies, insurance eligibility, payroll information (e.g., year-to-date FICA paid)
- order entry
- inventory queries and stock availability
- educational institutions: transcripts, grades, and course registration
- sales results
- payables status
- investor relations: stock prices, dividend reinvestment plans, major announcements
- general bulletin board
- Help Desk: automated answers to most common questions
- survey automation
- fax-on-demand (request specific fax information based on an indexed menu)
- health care, test results, physician referral

Many of these applications are available from vendors as industry tailored, off-the-shelf products. For example, the Fijitsu *Intervoice* IVR offers packages for employee benefits, bank processing, credit union, bill payment, 401K, and health care.

Call Center Applications

IVR is the *sine qua non* of call centers — they live or die by the efficiency and customer satisfaction provided by their voice response systems (as well as their CTI applications).

Call centers are the drivers of telephony innovation for large-scale IVR systems. These centers not only need more automated tools than other businesses but have a plethora of reporting capabilities that allow them to quantitatively justify the investment. Fortunately, many of the technology advances originating from high-end call centers are quickly promulgated to low-end IVR systems (some costing less than $2000).

Some of the applications found in call center IVRs include:

- Announcements. Callers can listen to a standard statement regarding products, services, etc. The IVR system can also use mathematical algorithms to announce to the caller an estimated wait time. Announcements can be tailored to the caller's interests; if they select option 3 to order fishing gear, for example, they can listen to fishing conditions on Norris Lake and the latest lures.
- Bulletin boards. These provide detailed information to callers on specific, selected topics.
- Leave message. Callers in the queue can elect to leave a message for callback, and then hang up. The first available agent calls back the customer. The callback can be scheduled or immediate.
- Supervisor observation. This allows supervisors to monitor an agent's performance.
- Advance information collection. While in queue or simply on the front end of a call, the IVR system can collect necessary information (e.g., account number or claim number) so that the agent has the information needed when the call is answered.
- Tracing calls. Calls are recorded for follow-up. This could be for malicious or emergency calls or simply sales leads. Also, some systems allow selective recording so that only the critical part of the conversation is recorded — usually the actual order or decision by the caller ("Yes, I would like ten blue shirts, 35×17, with the monogram XYZ").
- Conferencing. Agents can conference with another agent (e.g., a second-tier technical help engineer) while the customer is on the phone. Of course, the second or subsequent conferee need not be physically in the same building.
- Fax-on-demand. Callers can receive fax information either directly or while they are in the queue. Typically, fax-on-demand applications will attempt several times to deliver the fax (the number of tries is a system-level option).
- "Sticky" data. Information obtained in one part of the IVR system is passed to other applications so that the customer does not have to repeat the same information to another agent (having to repeat information is near the top of customer complaints about poorly designed IVR systems). To handle this process, the applications must have the ability to return to the original application (or script).

FAX-ON-DEMAND

Although fax-on-demand is merely another application like those listed above, its wide deployment in the IVR world merits a closer look. With approximately 100 million fax machines worldwide, it is an ideal medium to disperse information to customers, sales reps, etc. It is also easy to use and, unlike e-mail, always looks exactly the way it was sent.

There are several ways that the application can send the information. Using an index, which lists documents by a code (e.g., document #100 = claim form, document #101 = application for insurance, document #203 = engineering schematics for widget A, etc.), users use touch tone to indicate the document number (or say the number). More advanced systems allow the user to say the document; for example, "send me information on your Equity 4 mutual fund". Fax transmissions can be delayed until off-hours to reduce toll charges.

Fax-on-demand applications can enter variable information at the time of request. For example, benefits, wholesale electricity prices, mortgage rates, and account balance information can be dynamically updated (from the appropriate database) just prior to transmission. For many simple applications, the user can simply fax to a designated IVR port in order to store the document for later transmission to customers or other callers.

Broadcast fax can send to hundreds or thousands of recipients. For volume environments, attention needs to be paid to the fax database of recipients. The system can maintain a list of recipients who no longer wish to receive faxes, thus saving on long-distance charges. Cover pages can be tailored and retry options can be set to an appropriate value.

IVR APPLICATION DEVELOPMENT

Lucent's Script Builder and Voice@Work were discussed in an earlier section as a representative development environment. There are more than 100 vendors in the growing market for IVR development platforms. As DSP chip processing power grows (Moore's law of doubling processing power every 18 months continues to hold), packages to develop IVR and speech-related applications have become more user friendly, graphical, and effective. The programming legacy of reusable parts, abstraction from lower-level details, application generators, and use of templates thrives in the voice processing world.

To illustrate the details of IVR application development, two software packages — Apex Voice Communication's OmniVox and Pronexus's VBVoice — are presented below.

APPLICATION GENERATORS

OmniVox Applications Generator

Exhibit 3-3 shows a sample screen from Apex Voice Communication's OmniVox applications generator. It allows the developer to "drag and drop" icons reflecting call functions (Apex calls it a flowcharting tool). Various commands include call control, messaging, database access, faxing (see Exhibit 3-4), and outdialing. C functions (for the more technically inclined) are available for functions not provided in the GUI-based package.

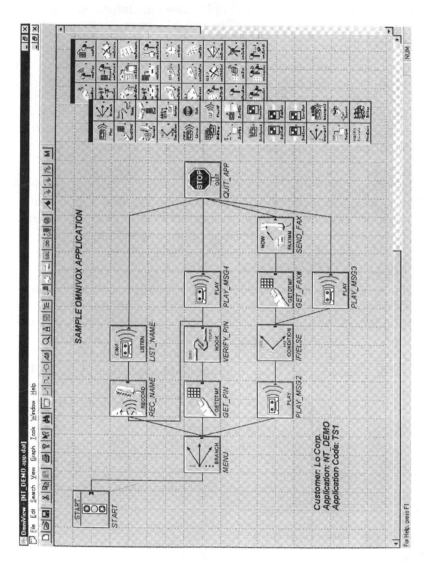

Exhibit 3-3. OmniVox applications generator.

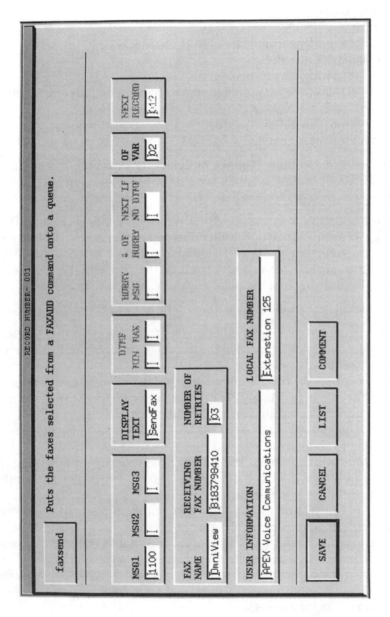

Exhibit 3-4. Adding faxing capability for an applications generator.

In addition to the flowcharting features, system utilities are needed for both development and daily operations. For Apex OmniVox, they include:

- start and stop line commands
- debugging mode
- error logs
- call counts reporting from call detail records (CDR)
- online display of all calls
- fax queue showing jobs to be sent
- fax append (allows several documents to be merged into one fax document)
- speech editor (can cut-and-paste)
- voice mailbox management

OmniVox runs on Pentium, Digital's AlphaServer, and VME-bus computers. Various UNIX operating systems are supported, including SCO, Open-Server, Unixware, Solaris, and Digital.

In contrast to OmniVox, Pronexus VBVoice runs on Windows 98 and NT. It is based on Microsoft's Visual Basic and relies on a series of specific telephony commands that are included on the left side of the VB screen. The developer selects the specific command, places it on the grid, and then connects them with a line drawing tool to create the application (see Exhibit 3-5).

VBVoice includes the following features (courtesy of Pronexus).

- 28 programmable telephony components that speed up the development cycle by rapidly allocating the low-level telecommunications operations within an application.
- Built in client/server functionality, with application messaging supported on local area networks and TCP/IP networks, including the Internet.
- Integration with any Windows application.
- Custom applications scalable to 192 lines of IVR on a single PC reduce overhead in Visual Basic shops.
- Native Windows drivers for the leading voice platforms, including Dialogic, Brooktrout, Pika, and Aculab, optimize performance and let the developer choose the best card for the job.
- Open TAPI integration to the office PBX and communication servers delivering call control to the desktop.
- Remote logging, operating as a separate executable, delivers a powerful maintenance tool to system administrators in the field
- Language rules support for up to 24 languages in the base package. Developers have a multitude of primary and secondary language options for their particular markets.
- High-speed access to ODBC data sources such as SQL Server and Oracle. The jet engine allows for fulfillment of enterprise telephony applications with easy access to legacy databases.

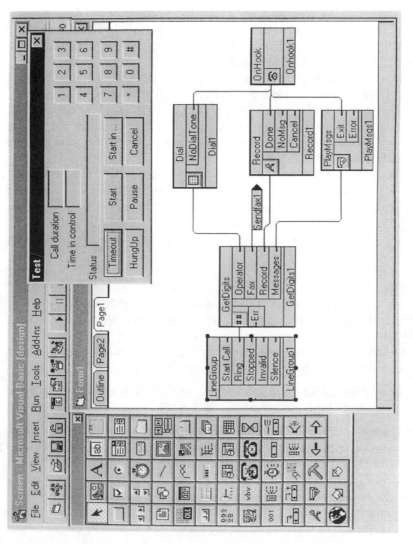

Exhibit 3-5. Pronexus VBVoice.

- Support for Windows multi-tasking ensures the stability of mission-critical applications.
- Support for all digital transmission protocols, including E1, T1, and ISDN.
- PBX and ACD-type functionality supports the ability to include call-transfer functionality without a PBX or KEY system, as well as "intelligent" call routing.
- Running on the V-12 and Daytona family of voice processing platforms from Pika Technologies, the POTS configuration will support up to 96 stations on a single PC and is suitable for any call management center with light to medium telecom traffic.
- Global error and digit handlers simplify the design for many applications. Detailed logs and status displays speed up the development cycle. Easy-to-use line handling and error logs make testing easy.
- A powerful property substitution mechanism transfers values between VBVoice components for easy access to stored property values like data fields.
- A built-in recording feature allows voice files to be recorded "on the fly" with a voice card or sound card. Pronexus recommends Announce! a full-featured Windows graphical voice recording and editing software.
- VBVoice supports distributed three-tier client/server applications and team-based, rapid application development for enterprise applications.
- Full conferencing and line bridging; conferencing for an unlimited number of participants on platforms scalable to 96 lines using SCBus architecture. (This feature and the next two are optional add-ons.)
- A large library of multilingual system phrases from Worldly Voices and a recording service in the service pack.
- Multi-line support for speech recognition (ASR) and dynamic conversion of text into speech through speech synthesis (TTS).

To illustrate the lower-level software components in a VBVoice IVR system, the following components are listed.

- CallQueue. Call management component that will manage incoming calls and distributed calls to agents. It contains many built-in features for ACD, including call routing, workflow methods, and priority number processing.
- AgentX. An ActiveX component that enables a communication link between peers along any TCP/IP network (LAN or internet). The component is used to exchange messages and dates or initiate disparate applications.
- VBVFrame. A container for all VBVoice controls. VBVFrame provides the palette on which components are connected and the projectwide properties and methods are set.
- CallTimer. Initiates a timer for setting process options like duration of a call or logging time in a process. Call flow is routed based on results

(e.g., time expires) and timer can be set to automatically expire, as required.

- Count. Maintains an internal counter, which can be reset or incremented. The call is transferred to a new control, depending on the results of a comparison between the counter value and a limit value. Useful for simple loops.
- DataChange. Changes data in a database record, in one or more fields. The database and record are pointed to by a previous DataFind control.
- DataFind. Searches for the first or the next database field, depending on whether the control is entered via the main input or the "next" input. A range of data match conditions is available.
- DataSwitch. Reads a database record from a previous DataFind control, stores the result in the property "Result," and routes according to the value obtained.
- Delay. Implements a wait period. The caller can be put on hold, or a greeting can be played during the wait.
- Dial. Dials some digits either during a call or to start a call. Can perform call supervision (answer, no answer, busy, etc.) and has built-in support for PBX call transfer.
- GetDigits. Plays a greeting and waits for digits from the caller. Configurable exit conditions using wild characters, numeric characters, and ranges. Built-in invalid digit and no digits handling.
- IniSwitch. Searches for a setting in a Windows initialization file. Branches to another control, depending on the value found.
- InConn/OutConn. Connect controls across forms.
- LineGroup. Channel manager. Waits for incoming ringing on a channel, or it can start a call. After a preset number of rings, the channel is taken off hook and control is passed to the next control. Analog and digital support available.
- OnHook. Plays a greeting, and hangs up the phone.
- PlayGreeting. Plays a VBVoice greeting. Provides options for fast forward, rewind, and pause.
- PlayMsgs. A high-level control that accesses a voice mailbox containing a list of messages. Configurable options after play. Supports new, old, and deleted messages. Updates database with new message status.
- Record. Plays a greeting, records a file, and adds a new record to a database. At the end of the recording, it allows the caller to choose options to re-record, delete, append to end of recording, or save the message.
- SubStart. A subroutine start control that connects the application to a new form. Useful for reusing blocks of logic.
- SubEnd. A subroutine end control used in conjunction with a SubStart control. When the call hits a subend control, processing returns to the initiating SubStart.
- TimeSwitch. Transfers the call to another control according to time of day and day of week.

- User. Allows the use of VB code to create controls. Access is provided to all the high-level VBVoice functions and low-level hardware-specific driver functions.

A final word on the development environment. The first contact an end user/customer has with an IVR system is the voice. If it is poor quality, there is an immediate negative impression — it is judged unprofessional. Thus, a critical part of the development package is a studio effects capability, including professional voice editing, music mixing utilities, sequential .WAV (or other format) compilation, and other tools of the voice trade.

BEYOND TOUCH TONE

Phase one of telephony automation is to use IVR in place of or along with human agents. Phase two is to use human speech to supplant touch tone as the *primary* IVR interface. Before DSP (digital signal processing) chips achieved the speed they have today, the IVR developer had two choices for automated speech recognition: (1) a few words, such as digits 0 to 9, that could be spoken by any caller and be recognized by the speech decoder, or (2) large vocabulary but confined to a single speaker (requiring training, as found in packages such as IBM ViaVoice). Although English has a foundation of only about 40 phonemes, the mathematical models required to convert speech into discrete words (using Markov algorithms, among others) are demanding indeed.

With faster chips and larger memories, today's IVRs can recognize large vocabularies from a diverse caller population. Ron Croen, CEO of Nuance Communications, uses the following example to illustrate the power and convenience of automated natural speech understanding.

With touch tone:

> Press 1 to buy shares, 2 to sell shares, or 3 for quotes
> "1"
> Enter the stock symbol
> "43936291"
> Enter the number of shares you want to buy
> "500"
> Press 1 for market price, or 2 to specify another price
> "1"

OR

> With natural speech:
> What would you like to do?
> "I want to buy 500 shares of Coca-Cola at market"

In addition to the translation of spoken word to text (digital) word, speech recognition also provides for speaker verification. Veritel, for example, offers a security product that compares the speaker's voice prints (for selected words) to a preregistered set of prints. If there is a match, the caller is allowed to access system resources, such as dial tone at an organization's PBX, with the ability to call long distance.

The following sections illustrate some of the applications of speech recognition in the marketplace. As home PCs begin to achieve chip speeds of 0.5 GHz, speech recognition capabilities in IVRs as well as many other devices will become *de rigueur.*

Voice Dialers

There is a lot to be said for doing one thing extremely well. Voice dialers do nothing more than connect one party to another via a spoken name. Lucent Technologies' Name Dialer© and Parlance's Name Connector© are examples of this technology.

Using Parlance's product as an example, here is how voice dialing works from the end-user's perspective.

- Dial a number (always the same number) that connects the user to the Parlance server.
- After the beep, speak the name of the person (or department, service, etc.) one wants to reach.
- The server repeats back the name — "if you stay on the line, you will be connected to John Doe."
- If the name is wrong, press "*" and say the name again; if it is right, just stay on the line.
- After a slight pause, one is connected to the party.

Parlance uses off-the-shelf Intel-based hardware, the NT operating system, and their own proprietary voice recognition system to make this work. The voice communications department supplies text-based names (no recording of end-user names required) that are translated by the Parlance engine into phonemes that are compared to the spoken voice. Usually, the text names can be downloaded from an existing directory or phone book. Nicknames are essential and are generated by the software for known common names (Dick for Richard, Bill for William, Janie for Juanita, etc.). However, Bubba, Slim, and other such nonstandard nicknames must be entered manually.

As of this writing, Parlance has a limit of 15,000 names (nicknames are not counted against the limit) for a single machine/directory. For very large organizations, cascading voice connection systems are required. For example, "West Coast" and "East Coast" servers could be set up. An employee working in the East Coast division could dial the East Coast server, say "West Coast," and be connected to the West Coast name connector. From there, the employee could say the name of any employee in the West Coast division.

The accuracy of voice dialing is approximately 90 to 95 percent. Users with soft voices, those whose native language is not English, and very fast speakers sometimes have lower hit rates using Parlance. Speaking on a cellular phone or over a speaker phone will also somewhat reduce accuracy. The Parlance name database was developed from both American and Britain speakers and will be accurate for those populations.

Despite its limited function, voice dialing is extremely popular with users once they grasp what it means — driving down the highway and being able to call or leave a voice mail for anyone in the organization by pressing one button on their cellular phone. Or not having to take the time to look up a number in the company telephone book. Or not looking up the number in the phone book, dialing the number, and finding out that the phone book is out of date.

Vendors selling these services tout the reduction of number of calls received by human telephone operators. And they imply that cost savings can be obtained by eventually reducing operator headcount. In practice, the greatest benefit is efficiency of the work force. For example, executives in their cars no longer need to call their administrative assistant and ask them to transfer to another party. Whether operator headcount can be reduced depends on a number of factors, such as the number of internal versus external callers, the aptitude of the work force for new technology, and the kinds of services the operators offer (simple transfer versus research on what department can best help the caller). One Houston-based energy company installed Parlance and found that the number of calls to the Name Connector servers exceeded the number of calls received by live operators.

A name connector service can be set up as a network outside the premises of the organization. Key customers, suppliers, government agencies, disaster recovery groups, etc. can be set up to speed access. Of course, the name "Dial Tone" with a number of "9," is not a good idea — hackers will find it quickly. Generally, any pronounceable name going to any specific telephone number (domestic or international) can be safely inserted into the database.

From an operational perspective, the following housekeeping functions are required to keep the name connector up-to-date and efficient.

- regular (weekly or monthly) feed from the organization's online or manual telephone directory to a text file that can be imported into Parlance's text directory
- routine examination of names that are difficult to pronounce or are pronounced in a manner that is not obvious from the spelling; phonetic spellings are entered into the database so that Parlance can recognize the name from the pronunciation (e.g., or example, "Fahim Bhakro" might be spelled phonetically as "Fa heem Bac row")

- backups of customer-specific files (to Iomega zip diskettes recommended by Parlance)
- digital port connection (the software must be set up beforehand for the specific PBX transfer function, which will vary by vendor)

Exhibit 3-6 shows a schematic diagram of the server connection to the PBX. Note that because of the simple analog T1 (tip or ring) connection, virtually any PBX can be connected to the server.

Lucent Technologies, through Bell Labs, has developed a package with capabilities similar to the Parlance Name Connector. The Large Vocabulary Recognizer, run on the Lucent Conversant IVR server, can store up to 20,000 names in a single directory. USAA in San Antonio, TX, has installed and used the Large Vocabulary Recognizer Dialer to help employees reach each other more easily and offload work from the internal telephone operators. In addition, they have plans to use the product to offer customers a more intelligent automated response. Rather than the traditional IVR, "For loans, press 1...," the savings bank will use the Conversant and Large Vocabulary Recognizer to respond to a question such as, "I need a loan."

IVR PERFORMANCE AND TUNING

Publilius Syrus, a first century B.C. Roman writer, admonished his readers to, "Never promise more than you can perform." It applies equally well at the beginning of the 21th century — underpowered IVR systems can look good in terms of features but fail to perform under the stress of large volumes. Even high-end IVR platforms need to be appropriately tuned to obtain a pleasant caller experience. The following are some tuning techniques that should be applied to new or existing IVR installations.

- Compression levels. More data can be stored on hard drives if compression is increased. This may mean a lower sampling rate and a corresponding decrease in the quality of prerecorded messages. Voice data is often stored in PCM (pulse code modulation) format (8000 samples per second), but further compression using ADPCM (adaptive differential pulse code modulation) squeezes the data from 64Kbps to 32Kbps or less. At 32Kbps, the caller will rarely notice a degradation in voice quality. At 8Kbps, the stored messages are noticeably distorted in most systems. However, several vendors are developing more efficient coding schemes. For example, Cybernetics Infotech, Inc., advertises near-toll-quality compression at 4Kbps and toll quality at 6Kbps.
- Cache Memory (buffer). Many callers may listen to the same speech segment. It is more efficient to put that section of digitized speech in a buffer so that access does not require an I/O operation by the disk drive read heads. If the IVR allows this parameter to be tuned (i.e., it is an addressable parameter), it should be reviewed to ensure that it

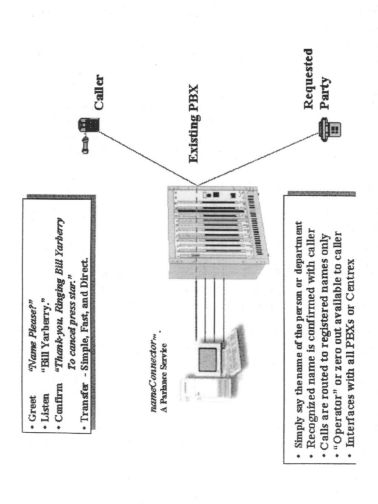

Caller

Existing PBX

Requested Party

- Greet   *"Name Please?"*
- Listen   "Bill Yarberry."
- Confirm   *"Thank-you. Ringing Bill Yarberry*
  To cancel press star."
- Transfer – Simple, Fast, and Direct.

nameConnector™
A Parlance Service

- Simply say the name of the person or department
- Recognized name is confirmed with caller
- Calls are routed to registered names only
- "Operator" or zero out available to caller
- Interfaces with all PBXs or Centrex

Exhibit 3-6. Parlance Name Connector: dialing by spoken name.

is adequate. Just having sufficient physical RAM will not guarantee adequate buffer space.

- Memory requirements. Memory is like riches — one can never have too much. Performance is greatly enhanced if most or all of the applications can reside in memory. It is particularly true for speech applications because they are realtime and callers expect quick response. If a sluggish system has to retrieve a voice file from disk, the caller might assume something is wrong with the system because the response is so slow.
- CPU utilization. Ideally, the CPU should be utilized in the 50 percent or less range. At 70 to 90 percent, some "thrashing" begins to occur; the system begins to spend considerable time managing overhead between all the transactions (swapping) and overall performance degrades.
- Phrase length. The IVR must perform additional operations if too many short phrases are used. An extreme example would be to have three phrases "your," "bank," "balance" rather than "your bank balance."
- Nonproduction activities. Activities that are not directly related to serving the caller should be monitored closely. Developers should either use another (presumably smaller) test IVR system or should test during "off-hours," if there are any. Logs, reports, and backups should be scheduled for off-peak hours.

A note of caution: IVR programmers sometimes do not talk directly to customers. Statistics coming out of the machine may not tell the whole story. Occasional surveys and other means to discuss customer satisfaction may indicate areas of poor design or — equally important — areas where IVR could be better integrated with other (CTI) applications.

IVR DESIGN

Much has been learned about good and bad IVR design over the last couple of decades. Some IVR applications have been astonishingly effective. For example, Len Dorrian, NationsBank senior vice president of strategic technology, reports that two-thirds of its calls never leave the IVR system (reported by *Teleconnect Magazine*, February 1998). Exhibit 3-7 contrasts practices that have been shown to be effective versus those that either drive customers away or unnecessarily increase agent workload.

Callers will retaliate if an IVR system is poorly designed. They will opt out to an operator, call management directly and complain, hang up in the middle of a session, or worst of all, simply go to another firm for services.

SIZING THE IVR SYSTEM

Ordering the correct number of ports, DSP cards, memory, hard disk capacity, and other components is not achieved by straightforward calculation — particularly when the system is new, the organization does not

Exhibit 3-7. Effective practices vs. counter-productive practices.

Productive	Counter-productive
Monitor blockage rates and busy hours	Too many options at one time (more than 5)
Simplify menu structure	No capacity to correct errors
Use voice recognition for rotary phone callers	Caller gets stuck in lower-level menu and cannot return to main menu
Install sufficient incoming lines to avoid busy signals	No way to reach a live agent
Ensure that agents' utilization rate is not so high that they become irritable	Use jargon or complex wording, confusing the caller
If the business case justifies it, develop an expert system that will guide callers through a solution to a problem (even if only 30 percent of callers can be helped by this technique, it takes a significant load off human agents)	Use inconsistent touchtone sequences to return to the main menu
Log where callers make errors and redesign	Set up servers, lines, and databases to handle the "average" load rather than peak load
Use messages that assure callers that system is not hung up while they are waiting. For example, have the IVR system say "Please wait, while we verify your policy number..."	Use inadequate error checking. Callers will, by accident or confusion, enter virtually any sequence of touchtones.
Use efficient algorithms for database access (e.g., indexed rather than sequential access)	Use error codes that are not meaningful to the user ("error code A345, call terminated")
Say the function first, the number second. For example, "For sales, press 1; for service, press 2" — not "Press 1 for sales, press 2 for service."	Avoiding feedback after the caller enters a menu choice
Combine IVR with Web applications (such as Edify's Workforce Server) to provide maximum servers to the customer	Do not provide repeat callers the option to avoid lengthy menus if they know the exact sequence of touchtones to enter
	Excessive depth of menus and too many options

know the volume of callers, how long they will stay on the line, how many long versus short messages they will leave, and how many of them will opt out to a live agent.

One approach is to use the iterative method — order an initial configuration, try it for a few weeks, look at the logs, and assess the adequacy of the system. A high rate of abandoned calls and a long queue length are good indicators of an undersized system. If significant increases in volumes are expected, the platform selected should be expandable and each module should be seamlessly integrated into an enterprise network. The prudent manager will build a contingency clause into the purchase contract so that the organization is not "stuck" with an undersized box should initial volume projections prove too low.

Some capacity-related options (in addition to the components mentioned earlier) include:

- Number of ports. Both voice and fax ports (if applicable) must be specified. The voice ports can be divided into digital and analog (tip or ring). If an organization intends to change PBX vendors but not before purchasing an IVR system, then analog ports may be preferable because they will adapt to virtually all PBXs. Later, when the new PBX is installed, specific digital cards or ports can be inserted into the chassis (if the IVR has a digital link to the PBX in question).
- Amount of voice storage. Hours of storage will vary by quantity of stored and received messages. Redundancy will increase the disk space required for the same quantity of voice storage.
- Interface options. These include T1 digital trunks, analog line cards (with options such as loop-start, ground-start, DID, and E&M for direct switch links).
- Number of serial ports and maintenance modems.

REPORTING AND REMOTE MONITORING

IVR applications, if they are well designed, quickly become an essential part of the business. They become "production" and thus must be monitored and proactively maintained. For performance and trending, IVR reports provide essential information.

As with other communications devices, IVR hardware should have a LAN interface and an IP address that allows SNMP management and monitoring. If a communications package such as HP OpenView, NetView, etc., is installed somewhere within the organization's network, the IVR hardware and software status can be monitored remotely.

Call statistics are vital for on-the-fly calling trends. Following are some example reporting features found on many IVR systems.

- call detail — how many calls and minutes both incoming and outgoing (for fax)
- daily, weekly, monthly reporting
- utilization by menu items
- time of day analysis
- fax utilization
- on-screen viewing option (or browser-based option)
- specialized reports by customer-entered account numbers, etc.; report wizards allow custom reports to be developed as needed
- ability to obtain reports while the IVR is still processing calls and faxes

SUMMARY

Interactive Voice Response continues to play an important role in virtually all areas of public and private commerce. As development tools become more graphical and utilize more reusable objects, the pace of development will continue to accelerate. Smaller and cheaper platforms will house IVR, and those platforms will soon go beyond anything now contemplated. Several years ago, AT&T predicted that by the year 2006 machines capable of routinely translating languages "on the fly" and in native voices would be available. With the doubling of processing power every 18 months, that prediction seems reasonable. It also implies a vast potential for IVR to provide 7 × 24 services, improve accuracy, and reduce costs for both domestic and international organizations.

Chapter 4
Fundamentals of CTI

COMPUTER TELEPHONY INTEGRATION (CTI) IS THE SET OF SOFTWARE AND HARDWARE COMPONENTS that allows a computer to manage telephone calls and integrate telephony services into desktop computers, servers, PBX devices, and other computing equipment. The generic term "CTI" is often used to include IVR, unified messaging, and sometimes voice over IP (VoIP).

CTI is growing at roughly 30 percent per year, with the small office–home office (SOHO) segment accounting for a significant part of that growth. This chapter focuses on the technological underpinnings of the myriad of applications that are being developed for organizations of all sizes. Later in the chapter, examples of CTI implementations will be described.

Viewed in the perspective of several decades, the drive toward CTI is similar to the "escape" of the user from the straightjacket of the mainframe in the 1980s. Using personal computers, far more computing power was brought to the knowledge workers. CTI is equally revolutionary; the power in servers and the PBX is now being distributed to many more people than before. The value of telephony is finally being fully realized.

GENERAL FUNCTIONS OF CTI

Later sections in this chapter will address CT functions at a detailed level. From a high-level perspective, CTI provides the following:

- First-party call control. Using a PC and modem, a telephone or telephone-like instrument can make calls based on a database or directly entered phone number. First party allows calls to be controlled from a single computer but not calls going to others in the organization (e.g., to reroute other agent's work when they are on vacation).
- Third-party call control. Ability to view and control calls coming to others in the organization (e.g., pick a call ringing on another person's extension).
- PC as telephone. With appropriate circuit boards (e.g., Dialogic boards), the PC can be made into an ersatz telephone, with the standard functions shown as graphical icons. This is often a SOHO (small office–home office) solution.
- Automatic recognition of incoming calls based on caller ID or ANI. Use of that information to retrieve or manipulate information on databases.

57

- Automatic logging of call activity (via logs with date and time stamps). Like messages (fax, e-mail, and voice mail) associated with a specific customer can be stored together and indexed to provide a complete record of communications with the customer.
- Fax management.
- A series of voice-activated services, such as text to speech or speech to text.
- Display information about the call on the monitor. Using CTI, previous calls and the current call can be shown in far more detail than is available with the limited LCD display space found on most telephones.
- Multiple call handling facilitated via a computer screen interface.
- Easy setup of multiparty calls.
- Using Internet tools such as JTAPI, browsers can be telephony enabled so that customers and employees can use online directories and complete transactions using voice communications after reviewing information on the Internet.
- Allows for innovative functions such as tracking down key employees in an emergency (e.g., call first goes to office number, then cellular phone, then pager, then next on the emergency notification list, etc.). Personal agents can read the time, date, and topic of a meeting over the telephone to the appropriate party.

BASIC ARCHITECTURE

The three common configurations for computer telephony integration include:

- Client/server. The connection between the telephone and the desktop is a logical, not physical connection. Telephony connections and functions are implemented via a telephony server. This architecture is sometimes referred to as third party. See Exhibit 4-1.
- Desktop. There is a direct, physical link between telephone and the desktop. Special hardware is required to ensure that the proper DTMF signals, etc. can be sent to the PBX from the computer. This first-party configuration is maintenance intensive for large organizations. See Exhibit 4-2.
- PC as telephone. With the appropriate telephony boards and multimedia hardware, the PC can function as a telephone. Of course, when the LAN or network goes down, the user has no telephone. Given the problems in most organizations' legacy wiring infrastructure, this option has not been widely deployed for critical business functions.

APIS AND CT STANDARDS

CTI would be impractical without application programming interfaces (APIs). The code to control telephony via the computer would be immense

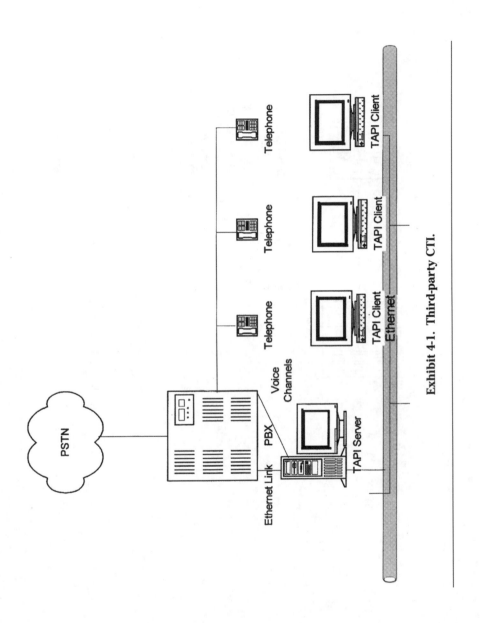

Exhibit 4-1. Third-party CTI.

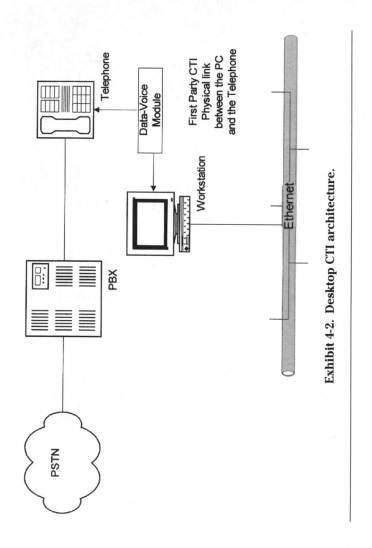

Exhibit 4-2. Desktop CTI architecture.

and would vary for each platform. The developer needs to be able to send commands to a software layer that abstracts the hardware layer from the application. APIs also allow code to be ported (with a reasonable level of effort) to older hardware, thus saving the organization considerable expense. Lucent Technologies (www.lucent.com), in a white paper, "Introduction to Computer Telephony," notes:

> For developers working in Windows NT or Windows 95, the RealCT API simplifies the process of creating CT applications. RealCT encapsulates several API functions in a single function call. For example, the process of placing a call with call progress monitoring for all Lucent hardware can take more than 300 lines of code. Using RealCT, the entire process requires only one function call. RealCT also manages all resources in the system, so your application does not need to keep track of resource use...

APIs are used by a program to accomplish a specific activity. For example, a programmer could write a simple dialer in Visual Basic or Java that would dial any number highlighted (the user can keep a name and address list in MS Word and invoke a VB program that dials the highlighted number via a modem connection). With the growing use of call controls from the desktop, a standard interface is essential to allow developers a straightforward way to provide users with all the telephony options available on the machine.

The three APIs in commercial use today are TAPI (telephony API), TSAPI (telephony server API), and JTAPI (Java telephony API). Each architecture has significant differences and should be reviewed carefully before a major telephony development is initiated.

TSAPI

TSAPI was the first API and was developed by Novell and AT&T (Lucent). It is designed to integrate the PBX with Novell's Netware LAN. Major functions include call control, call routing, monitoring, directory services, and query. It is server rather than client based and requires a Netware file server to be physically linked to the PBX.

TSAPI provides good call control because of the logical link between the telephone and the workstation. Server-based applications are called third party and have the advantage that no physical link is required between the PC and the workstation and the telephone (as is the case, for example, with Siemens COM Manager). This makes TSAPI less costly to implement in a large office where hundreds or thousands of legacy telephone sets may be in place. Another advantage to TSAPI is that many operating systems are supported (e.g., Mac OS, OS/2, UNIX).

At one point, TSAPI seemed to be languishing in the marketplace, but recent efforts by Lucent to continue development may lengthen its life. It appeals to large companies that may have multiple legacy operating systems.

61

For example, Lucent has partnered with CCOM to support Phoneline, an application that allows the user to dial numbers, conference, and perform many functions on the PC that would normally have to be done on a handset. Phoneline uses TSAPI to drive directory lookup, logging of calls, and quick call setup. The telephony work is done at the server level, not at the PC. TSAPI has been a favorite of call center developers because of the rich call control feature set (e.g., support of agents, skill-based routing, groups of interest). The biggest threat to TSAPI appears to be TAPI, which is supported by Microsoft.

TAPI

TAPI is delivered with Windows NT Server and NT Workstation, and is included in the Windows Open Services Architecture (WOSA). In addition, Windows 98 (client) includes a version of TAPI.

When first introduced, this first-party API was a rather limited function set of telephony capabilities. Its intent was to abstract the hardware layer so that developers could create software not dependent on specific hardware (device) configurations. Microsoft has steadily enhanced the capabilities to include support for IP telephony (i.e., the application using TAPI 3.0 does not have to know if it is connecting to the PSTN or an IP network) as well as Intel's universal serial bus (USB). TAPI is supported by more than 100 CT vendors and appears to be gathering momentum.

One of the obvious appeals of TAPI is the large installed base of the Windows operating system. There are two interfaces for TAPI: (1) software developers write code for the developer API, and (2) service providers such as PBXs, ISDN, analog phone system, and Centrex that must communicate through the SPI (service provider interface).

Individuals using Windows 98 are exposed to TAPI via the phone dialer in the accessories folder and the fax application (receives incoming faxes and displays them in graphical form). TAPI under Windows 98 can perform some other telephony tasks, such as answering incoming phone calls, forwarding calls, and proving voice mail (sold on some systems as "ring central fax").

TAPI also includes ActiveX support, which allows developers to use a graphical environment to develop modular programs that can be assembled into robust applications (ActiveX will be discussed later in depth). With the addition of object-oriented tools, TAPI will be a strong competitor for market dominance.

JTAPI

Java telephony API is a portable (i.e., can be ported to various operating systems), object-oriented application programming interface for Java-based computer telephony applications. Introduced in 1996 by Sun Microsystems in collaboration with a number of other telecommunications firms, JTAPI supports first- and third-party development. A Java virtual machine (VM) and a JTAPI-compliant telephony subsystem are the only requirements. Browsers such as Netscape and Internet Explorer are delivered with Java virtual machines. As a result, Web-based applications using JTAPI can run on the majority of platforms, using embedded HTML applets.

JTAPI is perceived to be more open than the other APIs. Indeed, packages such as IBM CallPath, TAPI, and SunXTL (for the SPARC/Solaris environment) can invoke JTAPI modules for telephony functions. Another strength of this young product is its portability — developers can move from one platform to another with little modification to their code.

The Java telephony model includes the common "objects" that are found in real phone calls — ringing, connection, addressing, termination, etc. JTAPI can interface with a telephony card inside a PC as easily as the traditional PBX. Addresses are, of course, telephone numbers.

JTAPI has some ambitious goals, including:

- simplify telephony development
- support both first- and third-party call control
- provide API for a wide variety of telephony applications and hardware or software platforms
- interface with existing TSAPI- and TAPI-based applications
- support more complex functions such as call center routing, with extensions to the basic set of features
- supply reusable objects to speed development

JTAPI is seen in the market as an enabler of the "Internet phone" where Web access and telephony are combined.

As an example of JTAPI implementation, consider a typical inbound call center. One of the functions for a customer service group would be to receive calls directed by an intelligent ACD and use an intranet or the Internet to access the business application and get screen pops as they answer the caller. JTAPI programming (similar to Netscape plug-ins familiar to most home Internet users) would allow telephone functions (make a call, transfer a call, conference, etc.) in a frame within the browser. A browser-enabled agent can also use Java Media Framework and Java Speech to integrate audio and video, so that both the customer and the call center agent can see and talk to each other. Traditional call center statistics (number in the queue, average wait time, etc.) could be programmed to display on the agent's browser screen.

Similarly, CT using JTAPI (or ActiveX, for that matter) can help outbound call centers. Dialers can be CT-enabled and fall into the following categories:

63

- Call preview dialers. CT obtains database records on a specific customer *before* the call is placed so that the agent is fully informed before speaking to the customer. This would typically be implemented for high value customers (e.g., more than $3000 business annually for a retail customer).
- Predictive dialers. CT does all the hard work of dialing, listening (via tones) for voice mail or ring no answer, and other conditions that would preclude a live conversation, then turns over — most of the time — a "live" customer to the agent. Screen pops are also used sometimes for prospective customers (e.g., to go against demographic data and judge the approximate family income of the prospective customer by the zip code or address).

Using an Internet connection, telecommuters can participate in the full functionality of an office-based communications system while still working from home or a remote office. The organization benefits from lower operating costs (in major cities, the cost of office space is frequently $5000 to $10,000 per person per year), extended hour coverage, more robust or quicker response to central disaster occurrences, and better work force management. JTAPI applications via an ISDN line provide the functionality over a wide range of equipment; the telecommuting agents can answer calls, transfer calls, and participate in ACD routing as appropriate to their skill set.

Developers in Java use the Java Media Framework API (JMF) to synchronize and control time-based data such as audio and video from within a Java application or applet. On the surface, it is analogous to the popular "real audio player" familiar to home Internet users who listen to their favorite stations over the Internet (via packetization of music). However, this "Java media player" controls audio, video, and MIDI across all Java-enabled platforms. JMF will support new CODECs. One of Sun Microsystem's goals for JMF is to support as many media types as possible and make JMF function on a wide variety of hardware platforms. JMF "players" can be written in the Java programming language.

JMF includes the following audio formats:

- AIFF (audio interchange file format)
- AU (UNIX file format)
- DVI (digital video interactive)
- G.723 (this is an important standard; it is the ITU-T recommendation for compressed audio over standard POTS lines)
- GSM (standard for cellular compression)
- IMA4 (interactive multimedia association standard for multimedia systems)
- MPEG (moving picture experts group — standard for reduction of storage requirements for video)

Exhibit 4-3. Comparison of the major computer telephony APIs.

API	Advantage	Disadvantage
TSAPI	• Most mature product • Supports complex routing and other functions needed by call centers • Better for environment with hardware/software diversity	• Novell has dropped support • Third party is only option (may not be economical for small business)
TAPI	• Most popular by far • Suitable for small companies (direct connection between phone and computer) • Support of Microsoft • Continuing rapid enhancements	• Not yet robust enough to support complex call center applications (although this may change soon). • Requires Windows-based platform
JTAPI	• More open • Has more features than TAPI • Focus on reusable objects and object libraries	• Has not had the same level of public exposure as TAPI. Newer releases of TAPI could become "standard."

- PCM (pulse code modulation — a common algorithm for converting an analog voice signal into a digital representation)
- RMF (rich music format — allows high quality music and sounds to be delivered over the Internet or IP circuits)
- WAV (well-established digital representation of sound waves)

JMF 2.0 will support sound and video capture. Organizations that need to disseminate audio or video information should consider standards such as JMF when developing their multimedia infrastructure.

The pros and cons of TSAPI, TAPI, and JTAPI are summarized in Exhibit 4-3.

ActiveX

ActiveX is part of Microsoft's component technology known as COM (component object model) and now DCOM (distributed COM for communication across networks). ActiveX allows developers to build dynamic telephony-enabled Web pages using software objects rather than coding detailed TAPI functions. When the end user views a page that has ActiveX controls, the components of the HTML page are downloaded at the user's hard drive and stay there for later use. The controls are objects created to perform specific telephony (or perhaps graphical) functions on the Internet, intranet, or telephony network. This is a convenient way to keep an application updated on each user's desktop.

ActiveX controls mesh well with the Windows operating system interface (as one would expect), but are also being ported to Macintosh and UNIX. As a result, ActiveX may further encroach on the market share of Java.

ActiveX operates with a different philosophy than Java. Java and Java-Beans run exclusively in the Java virtual machine (VM). This preserves security by preventing the Java executable from getting into the memory of other applications or executing low-level disk drive functions. This reduces some of the capability of Java (e.g., use of the right mouse button). On the other hand, Java works easily across multiple operating systems. Microsoft is attempting to address the security holes by promoting its Authenticode security system (Authenticode 2.0).

USING COMPONENT SOFTWARE

Writing code in Visual Basic, using TAPI to direct realtime telephony events, is for those who enjoy programming challenges and have lots of time. Real-world developers generally use packaged software components like ActiveX from Microsoft or JavaBeans from Sun Microsystems to insert telephony features into applications.

Productivity is greatly enhanced if realtime events can be handled by a package. The programmer must still understand what happens in a telephony environment (e.g., ring no answer), but the details can be handled by the component technology.

Java offers reusable components called beans. There are beans specifically designed for telephony that use JTAPI as their interface standard. Beans can be simple, such as a visual "slider bean" that allows the user to vary a field between 0 percent and 100 percent or a calendar bean, which provides data information. Like predecessor packages such as Visual Basic, Powerbuilder, or Delphi, JavaBeans can be assembled to construct applications (including telephony applications).

Using ActiveX, for example, telephony functions can be set up via drag-and-drop of reusable components in an application like MS Access or Visual Basic. Using these components, the developer can answer calls, select caller data from an IVR repository, generate voice prompts, and initiate screen pops. It is likely that as the graphical, reusable components become easier to assemble and debug, that the industry will move toward more customization of software rather than the limited-choice, off-the-shelf, hard-to-change software previously available.

DISTRIBUTED VERSUS DESKTOP CT

As CTI first began to be deployed in large organizations, the solutions (such as TAPI) were desktop-oriented. Telephony boards had to be purchased for each client and upgrades were expensive because technician had to "touch" each workstation. Also, telephony boards placed in workstations (such as a fax board) cannot be shared.

The solution that large organizations are moving toward is client/server (distributed) telephony. By placing telephony functions on a server, clients can move to a software-only environment (no local telephony hardware). Advantages to this approach include:

- The bulk of the maintenance is at the server rather than each workstation.
- Telephony resources can be shared throughout the organization.
- Upgrades are greatly simplified.
- Enterprisewide cost is significantly reduced. Since resources are shared, economies of scale necessarily obtain.
- Ports, boards, and other capacity-sensitive resources can be added to the server without affecting the desktop.

INTEROPERABILITY STANDARDS

The telephony world tended to be proprietary and closed until the 1990s. As applications became more complex and a much larger number of CTI vendors entered the field (as a result of increased user demand), the ability for hardware and software to interoperate became more important.

A full discussion of telephony standards would require many tomes. Some of the more important and more visible standards are discussed below.

MVIP

Multi-vendor integration protocol (MVIP or MVIP-90) has become the standard for integrating heterogeneous technologies such as telephone circuit boards. Within a single hardware platform (e.g., a PC chassis), multiple vendor cards can co-exist as long as they conform to the MVIP standard. MVIP supports 32Mbps throughput and 256 full-duplex paths on which data can flow (this should be adequate for even enterprise-level CT). In place since 1990, MVIP has been widely adopted for voice, fax, data, and video services by more than 170 vendors manufacturing board-level MVIP products.

Consider the hardware requirements for a large audio conference or traffic originating from a T3 or SONET link. MVIP ensures that there will be adequate capacity to handle all the traffic within a single-chassis or across multiple-chassis nodes.

H.100

Part of a group of interoperability agreements, H.100 specifies hardware configurations at the chassis card slot for a CT bus interface. It allows the industry to concentrate on the more pressing issue of software compatibility by reducing or eliminating incompatible buses. H.100 supplants both the competing MVIP-90 and SCbus standard. The specification was developed by an ECTF (Enterprise Computer Telephony Forum) Working Group.

S.100

Without voice processing software standards, multiple (different vendor) applications could not run on the same physical box. S.100, also developed by ECTF, specifies how CT applications can be developed in an open environment, independent of supporting software. For example, an organization may want to run fax-on-demand and text-to-speech applications from different vendors on the same NT server. If both are S.100, then — at least in theory — they should be able to co-exist on the same box.

H.323

Voice and other time-sensitive traffic is rapidly moving to packet-based networks. Anticipating this trend, the ITU-T developed a standard in 1996 that specifies how call control, channel setup, and CODEC specifications should work for realtime traffic (i.e., traffic that does not have a guaranteed quality of service to ensure that the flow of media is smooth).

The H.323 Protocol supports ITU G.711 and G.723 audio standards. It also supports the Internet Engineering Task Force's specifications for controlling audio flow to improve voice quality.

Microsoft's NetMeeting 2.0 is an example of an application that uses the H.323 Protocol. NetMeeting enables realtime, point-to-point audioconferencing over the Internet or corporate intranet. It includes features such as half- and full-duplex audio support for real-time conversations, automatic microphone sensitivity-level settings, use of MMX-enabled voice CODECs to improve compression performance, and microphone muting. By operating under the umbrella of the H.323 standard, this package can interoperate with other H.323 audioconferencing, such as Netscape Conference. NetMeeting also includes functions beyond audio, such as whiteboarding, videoconferencing, file transfer, shared clipboard, and chat.

LDAP

LDAP (Lightweight Directory Access Protocol) is often viewed more as a data than a telephony standard. However, because it houses the "white pages" for an organization (i.e., it is a directory service), it becomes important for telephony addressing as well. In other words, John Doe might have an entry that includes not only his Internet ID and building room number, but also his telephone number and fax number. LDAP can function as a directory on its own or can link to a full-featured X.500-compatible directory service. LDAP runs over TCP/IP networks. LDAP is relatively broad in its scope of addressing. For example, it can also store JPEG photographs, μ-law encoded sounds, URLs, and PGP (pretty good privacy) keys.

DEVELOP VERSUS BUY

IT organizations have always faced the "make or buy" decision. With the introduction of JavaBeans and ActiveX development tools, it is much easier now for organizations to telephony enable existing applications rather than having to buy a product for every need. This both reduces cost and increases application flexibility. Suppose, for example, a company has an intranet "find an employee" application. It would be far better for the users to telephony enable that application to allow the user to dial the person found via the search engine. ActiveX interfaces nicely with Internet Explorer and can produce the functionality relatively quickly. Otherwise, a separate package must be purchased (usually on a per-seat basis) and installed on an already crowded desktop (in terms of memory and other resources). It also means yet another package for users to learn.

Microsoft has developed an architecture called WOSA (Windows Open Services Architecture). Using WOSA allows programmers to write a set of API calls that can be used to telephony enable potentially thousands of applications that are now WOSA compliant.

APPLICATION GENERATORS AND CT ARCHITECTURE

One method to quickly generate telephony-enabled applications is to use an application generator, such as DavaViews DV-Centro. The best application generators will have the following characteristics.

- generate code in a standard graphical and object oriented language such as Visual C++ (code generation should work both ways — code can be generated from graphical design and graphical design should be reconfigured when code changes are imported back into the design)
- link application to Web pages via ActiveX, Java, or similar architectures
- interface graphically with the developer
- maintain an object library for real-world telephony functions
- limit or eliminate traditional line-by-line coding
- generate both source and executable code

CTI architecture is a serious, long-term decision for an organization. Sales presentations and demos, unfortunately, do not reveal all the key elements that will make a platform successful for an organization (whether an end-user organization or a systems developer). Developing and long-term maintenance are two sides to the same coin and should be carefully considered or costs will escalate over time.

There are somewhere between 50 and 100 application generators on the market now. The following are some characteristics that should be considered when selecting an architecture.

- True client/server architecture is essential. Without it, systems cannot be scaled.

69

- Portability requires Java or a Java-like script language. In addition, the application generator must be able to address various PBX and network standards such as E1 and ISDN, function in different languages, and access a plethora of databases.
- The GUI should provide help for new developers but not restrict the work of experienced developers.
- Like Visual Basic project files, all elements of an applications should be tagged and saved as a unit.
- The script language should be easy to learn and readily adaptable to the unpredictable nature of caller interactions. Graphical flow of call processing must be obtainable from script code.
- Documentation generation must be automated and part of the package.
- Modular architecture for CTI applications, as was true even for COBOL mainframe programming, is essential for effective long-term maintenance. As technology changes, only certain modules need change — not the entire application.
- The hardware platform should have no single point of failure — or least have the capability of functioning that way — if such robustness is required. For example, if all switching is concentrated on a single component and it fails, the entire system fails. For critical business applications, this is not acceptable. Non-stop/redundant processing (via "hot swappable" hardware) also supports 7×24 operations.
- Functions such as text-to-speech and database lookup should function independently, so as to increase overall processing efficiency and allow programmers to independently add, change, or delete modules as needed.
- Full Web enablement is essential. Resources such as telephony, fax, and access to databases should be available via a Web browser.
- System monitoring via SNMP to a monitoring package such as HP OpenView or IBM NetView is necessary for remote monitoring.
- Simulation tools, which allow component testing without live calls, are an important consideration. The CT package should mesh with the hardware testers available on the market.

MIDDLEWARE EXAMPLE

The generic definition of middleware is a software package that sits between one system (client) and another system (server), and facilitates the exchange of information between the two. To make this rather fuzzy definition more clear, the following subsection describes an illustrative middleware product: FastCall by Spanlink Communications.

An Example Middleware Product: FastCall

FastCall is a graphical development product that adds turnkey CT capabilities to Windows-based programs (including Help Desk software, personal information managers, and contact managers). It supports telephony enablement via TAPI and TSAPI APIs. According to Spanlink (www.spanlink.com), "FastCall is easily installed with little or no training and requires no customer programming or set-up costs. FastCall advanced CTI functions include inbound screen pops, intelligent call routing and call screening, coordinated voice and data transfers, outbound preview dialing, and robust screen-based telephony." The critical eye of telecommunications managers that have implemented real-world CTI systems may take issue with the words "easily installed with little or no training…" However, the description is relative — compared to straight TAPI calls out of Visual Basic or Visual C++, middleware *is* simple to implement.

The following description of FastCall is based on a Spanlink white paper available on their Web page. Other firms, such as Aurora Systems, Pronexus, and Teledata Solutions, have competing middleware products.

FastCall's telephony enablement of the Remedy Help Desk system will serve to illustrate the process and features of a middleware package. Remedy is a well-established Help Desk product that automates the tracking and reporting of problem calls (plus many other functions not addressed here).

In a typical Remedy implementation, an ACD routes the call to a Help Desk agent's desk, either via ANI or through the use of an IVR. When Remedy is telephony enabled, screen pops show caller information (from the Remedy database) as the call comes in. Without the use of middleware, programming a simple screen pop might take weeks or months of expensive programming. It would also require some modifications should the PBX be changed.

Exhibit 4-4 illustrates FastCall integration with the Remedy Action Request System, using the client/server model. The mechanics of the screen pop process is as follows.

- The PBX routes a call to an agent's station.
- The PBX also communicates with FastCall (on the CTI server).
- A DDE (dynamic data exchange) process is executed and a Remedy action request macro runs on the agent's desktop.
- Screens on the desktop are opened with relevant caller history, current information, etc.

The call processing flow at a more detailed level is as follows.

- Incoming call arrives. Calling number, called number, caller input, and time of call are identified as parameters.
- Call rules are checked (in priority order) against the incoming call parameters.
- A determination is made whether incoming call should be accepted (answered) or forwarded.

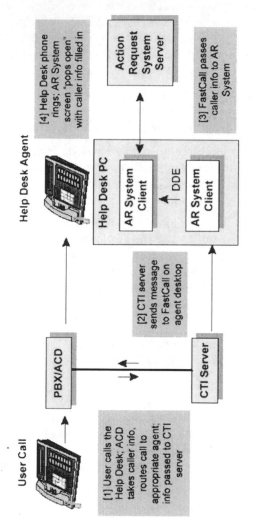

Exhibit 4-4. FastCall integration (client/server model).

- If incoming call is forwarded, determine if it is forwarded to attendant, voice mail, or a different extension.
- If incoming call is accepted, determine if any triggers (events that start a programmatic process) should be executed.
- Execute appropriate triggers for screen pop or other processes, based on incoming call parameters.

Number information is passed via ANI, DNIS (for toll-free numbers), or IVR response.

FastCall itself is a garden-variety Windows application (Windows 98 or Windows NT) and is stored in its own subdirectory and program group. It supports TSAPI and TAPI on the telephony side and DDE/keystroke macro interfaces for the computer.

Exhibit 4-5 is a screen print of call control keys that allow users to access telephony features from their workstation. The second major component of the software is shown in Exhibit 4-6. The administration programs provides for the logical flow of telephone calls. Call rules, view call activity, application paths (i.e., to reach executable subdirectories), and other preferences are included in this menu.

When FastCall is first installed, it must be tailored to the existing telephony and network/server environment. Exhibit 4-7 illustrates the screen that is used to specify telephony rules for the package. The generic process includes the following.

- Define the FastCall application paths that specify how to launch other Windows applications. This creates a list of application names (e.g., AR system, MS Word, etc.) that can be invoked when a call comes in.
- Define the FastCall application triggers that are the specific actions to be taken when a trigger is fired. Triggers are the specific functions in the other applications that are invoked by FastCall (e.g., invoke the AR system client and pass parameters to a submit macro).
- Define lists of calling numbers. These and caller input lists describe who might be calling, who they might be calling for, and the types of input the caller might give to an IVR. These lists are referenced by various FastCall processes.
- Define other set-up parameters, such as default area codes, extension formats, etc. These parameters help FastCall interpret telephone numbers.
- Define sets of incoming call rules that determine which application triggers to execute for different types of calls and call parameters. These rules provide the essential logic of processing.

In order to present a more in-depth picture of a CTI implementation, some of the more salient steps of the FastCall/Remedy installation are described

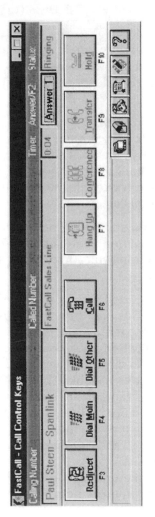

Exhibit 4-5. Call control keys.

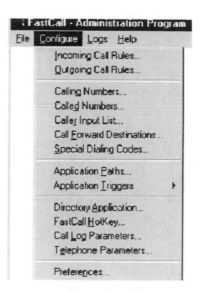

Exhibit 4-6. Administration program.

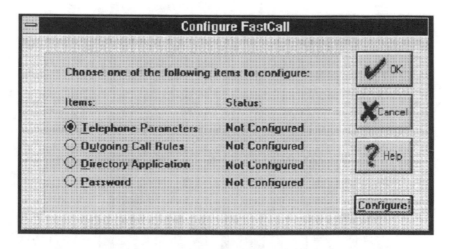

Exhibit 4-7. Specification of telephony rules.

below. (Although these steps may seem somewhat lengthy, they require considerably less effort than coding 5000 lines of TAPI calls in C++.)

- Flowchart the integration process (see Exhibit 4-8).
- Create an AR System macro (in Remedy) that records the result one wants to see when a call is routed to the agent desktop. Typically, this macro includes displaying a Query list of the open service requests for

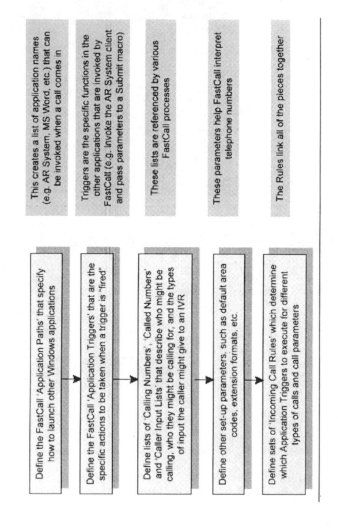

Define the FastCall 'Application Paths' that specify how to launch other Windows applications

This creates a list of application names (e.g. AR System, MS Word, etc.) that can be invoked when a call comes in

Define the FastCall 'Application Triggers' that are the specific actions to be taken when a trigger is "fired"

Triggers are the specific functions in the other applications that are invoked by FastCall (e.g. invoke the AR System client and pass parameters to a Submit macro)

Define lists of 'Calling Numbers', 'Called Numbers' and 'Caller Input Lists' that describe who might be calling, who they might be calling for, and the types of input the caller might give to an IVR

These lists are referenced by various FastCall processes

Define other set-up parameters, such as default area codes, extension formats, etc.

These parameters help FastCall interpret telephone numbers

Define sets of 'Incoming Call Rules' which determine which Application Triggers to execute for different types of calls and call parameters

The Rules link all of the pieces together

Exhibit 4-8. Flowchart the integration process.

the caller and a Submit window to enter a new request. The caller telephone number is recorded into the macro as a parameter that is passed from FastCall, and used as the key to identify the caller and do the query list display.

- Define a FastCall application path. This tells FastCall where to find the application to be run. See Exhibit 4-9.
- Specify an application trigger. This tells FastCall what process or function to execute when a call is being processed. Application triggers can send a DDE command to another application and can be defined for both incoming and outgoing calls. In the example shown in Exhibit 4-10, an incoming call trigger is added by clicking on the add button. The application trigger is given the name "POP" so that it can be referenced by the call rules to be defined later. The trigger runs the program labeled "ARSystemLaunch," which executes the program at C:\REMEDY\ARUSER.EXE. The trigger type specifies one of three possible values: Keystroke Macro, DDE Client Trigger, or DDE Server Trigger. FastCall can be either a DDE client telling another application what to do, or a DDE server responding to a command from another application.
- Configure the DDE client. If the configure button in Exhibit 4-11 is clicked, the client configuration screen appears (Exhibit 4-11). This screen specifies the details of the DDE command to run the Remedy AR system and execute the macro. The details of the DDE command structure are outside the scope of this book, but are well-documented by Microsoft and many third-party publishers. The data format field in Exhibit 4-12 contains the specific DDE command structure. The timeout is set to match the organization's network environment. Macro launching should be less than two seconds for most organizations.
- Define the telephone number structure. The formatting screen for a PBX telephone number scheme is shown in Exhibit 4-13. It is used to define how FastCall interprets telephone numbers passed from the switch. This screen is similar in function to the familiar Windows 98 dial-up networking screens (e.g., include or do not include area code).
- Set up incoming call rules. At this point, the rules that govern what actions are to be taken for incoming calls need to be specified. The previously mentioned application triggers are part of the actions if the qualification is true. There can be multiple incoming call rules defined; and for a given call, they are evaluated in priority order until one is found to be true. If an incoming call does match the criteria set up for a rule, FastCall will take the action selected in the "action desired" field (see Exhibit 4-14). Also, if a rule is true, FastCall will execute predefined alerts that have been defined under the alert button.

The net result of the above steps is that when a telephone call is routed to a particular desktop, the corresponding Remedy Help Desk AR system will pop open a set of windows with relevant information about the caller

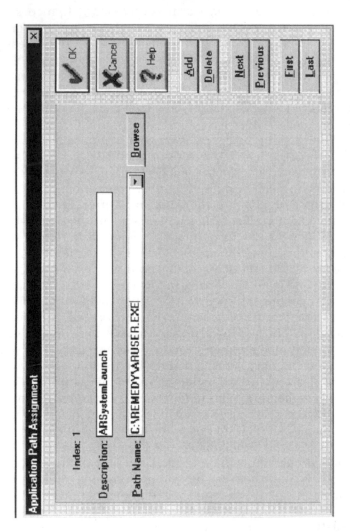

Exhibit 4-9. Define FastCall application path.

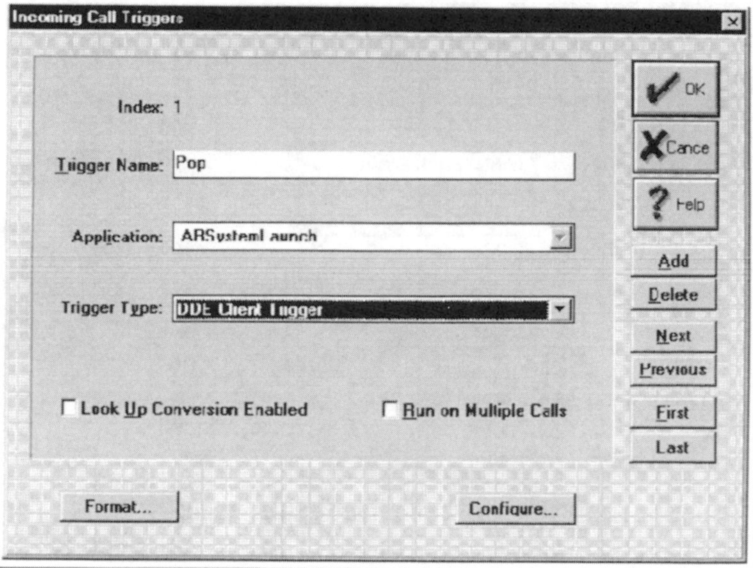

Exhibit 4-10. FastCall incoming call triggers configuration screen.

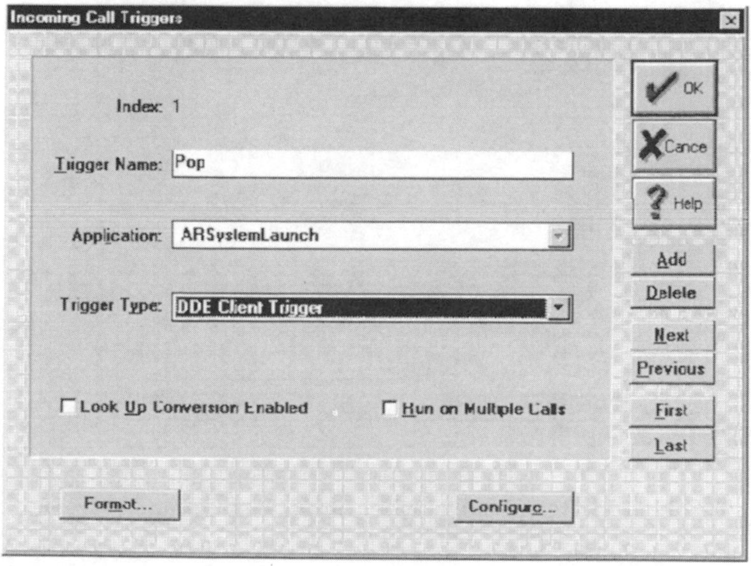

Exhibit 4-11. Configure DDE client configuration screen.

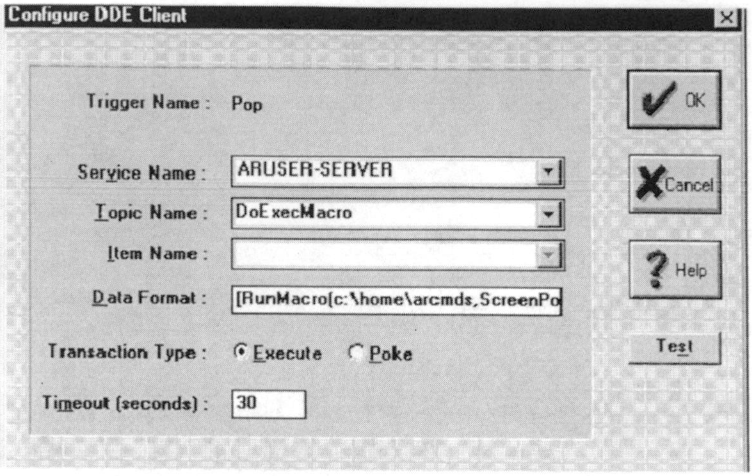

Exhibit 4-12. Configure DDE client.

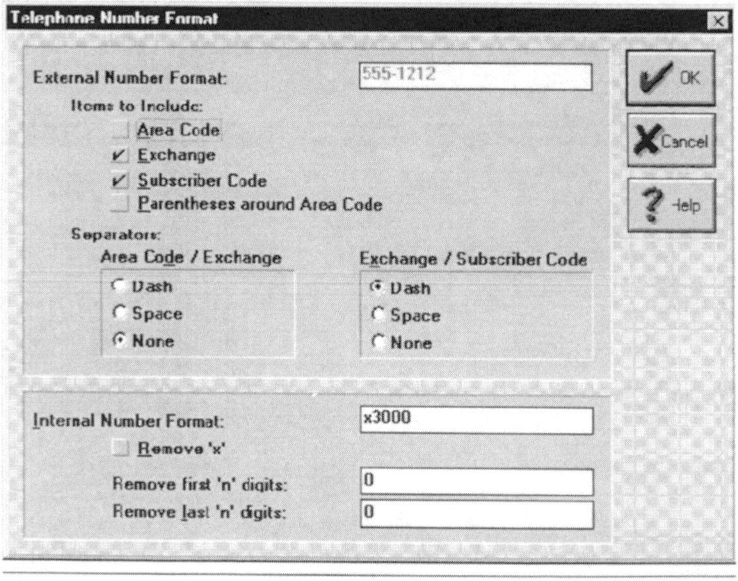

Exhibit 4-13. Telephone number format configuration screen.

(e.g., "Mr. Jones, I see you called last week about your hard drive; we will send out a technician within two hours...").

Middleware is clearly a benefit for organizations that want to telephony enable their existing applications using in-house resources. Although there

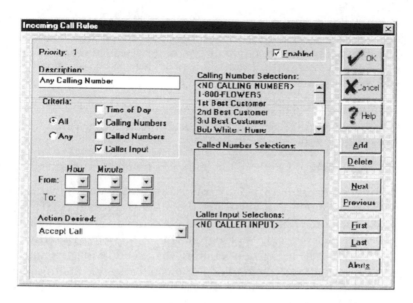

Exhibit 4-14. FastCall incoming call rules configuration screen.

may be niche applications that require lower-level programming, the majority of applications can be well served using middleware — not to mention the savings and significantly shortened completion time.

AUTOMATIC SPEECH RECOGNITION (ASR)

Moore's "law," stating that processor power will double every 18 months, continues to be upheld in the marketplace. Accordingly, speech recognition applications are rapidly being deployed in a variety of settings, including offices, homes, toys, and cars. Traditionally, there have been two approaches to ASR:

- Speaker-dependent systems. Using predefined text to capture the intonation, speed, and other voice characteristics of the speaker, the system is initially trained to a specific speaker only. Over time, the software improves its accuracy as mistakes are corrected. Speaker-dependent systems require less processing power and less complex algorithms to recognize grammar; they can thus afford a large recognition vocabulary. Examples of speaker-dependent speech-to-text packages include IBM Via Voice and Dragon Systems Naturally Speaking. Speaker-dependent voice recognition software has become so common that it is now more a commodity than a high-end platform.
- Speaker-independent systems. Using a generic vocabulary (usually generated by sampling a large number of individual speakers), speaker-independent systems recognize voice commands without special

training. For example, when the Parlance Name Connector package says, "If you wish to speak to an operator, say *operator*," its ASR engine listens for the most salient characteristics (in terms of an average voice print) of the word operator and acts accordingly if it is detected.

Recognition of the digits 0 to 9 has been available for years. Lucent Technologies' Conversant IVR platform, for example, includes a standard package for recognizing the ten spoken digits.

Recognition methods include the following common methods.

- Continuous recognition. The speaker does not pause between words and the system has to parse out the sounds into words. This is the most demanding requirement. Computers have a great deal of difficulty with certain sequences of words. Once, the writer Dorothy Parker, known for her keen sense of humor, was asked to make a pun on the word "horticulture." She replied, "You can lead a horticulture, but you can't make her think;" such a phrase would be most difficult for even the most discriminating ASR system.
- Discrete recognition. By slightly pausing between words, the ASR system can recognize words with a high level of accuracy. As processors increase in power and algorithms improve, this method will be relegated more toward the low end of the market because it takes less processor power than continuous recognition.
- Word spotting. The ASR system identifies key words and performs specific, programmed activities. For example, the caller might say, "How much does a VCR cost?" The words "cost" and "VCR" would be recognized and the caller would hear pricing on VCRs from a retail pricing database.

There are some common-sense guidelines that developers should use when designing an efficient ASR system, including:

- Avoid asking questions for which the response involves similar sounding names ("tree" and "three").
- Do not ask open-ended questions. The response from the caller should be a narrow range of choices, such as "yes," "no," "one hundred," or "operator."
- Avoid too much complexity in any single step.

OTHER EXAMPLES OF CTI APPLICATIONS

The following descriptions of demos from Nuance Corporation (www.nuance.com) illustrate the power that CT applications have developed as a result of better algorithms, faster processors, and newer development models.

Travel Plan Example

The Nuance Travel Plan demo shows how easy it is to get information about fares and schedules using speech. By simply interacting with the system in a normal voice, one can enter an itinerary and get realtime travel information. The demo includes information on 250 airports in the United States and abroad.

While trying the demo, notice these Nuance features:

- Speaker independence. For a system to be a success, it has to be able to understand a wide variety of regional and even international speakers. Nuance software does not need to be trained to recognize a voice, even if one is a non-native English speaker.
- Natural language. People refer to cities and airports by many different names. Because of this, a speech system has to be robust enough to handle wide variations not only in accents, but also in the way one refers to cities. For example, John F. Kennedy International Airport can also be called New York Kennedy, JFK, Kennedy International, and many other variations. Nuance builds these variations into the system so that it is easy to use. When selecting flights from the list offered, try saying phrases like "the second American flight," or "the seven thirty five flight."
- Barge-in. This demo allows one to speak when one wants to. At any time, one can respond to a question, even if the system is speaking. If already familiar with the system, barge-in enables even faster completion of transactions.
- Fast and easy to use. Getting flight information using touchtones is a slow and frustrating process. Even when one knows the flight number before the call, one still has to work through layers of menus and instructions to get the desired information. If the flight number is not known, one almost always ends up waiting on hold to speak to a customer service agent. With Nuance speech recognition, one gets the needed information quickly and easily.

Banking Example

Nuance Communications' Better Banking demonstration highlights the convenience of self-service banking. The demonstration incorporates personal account management, funds transfer, and bill payment functions.

Listen for these Nuance advantages:

- Natural grammars: the demonstration shows Nuance's ability to allow natural phrasing of dollar amounts, dates, and transaction types. For example:
  — Dollars can be "twenty-five dollars and thirty-two cents" or "twenty-five thirty-two."

- —Dates can be "June 3rd" or "June 3rd 1998" or "today" or "the first of the month."
- Natural language understanding: Nuance's sophisticated natural language processing accurately determines the correct meaning of naturally phrased commands such as "Transfer from Checking to Savings" or "Transfer three hundred dollars to Savings from Checking".
- Continuous speech: all Nuance software supports continuous speech. No artificial pauses are required.
- Accurate recognition: high accuracy and easy transaction modification are essential for user confidence. By combining superior core recognition technology and friendly interface design, one is confident that transactions will be submitted correctly.
- Speaker independence: the system does not need to be trained for your voice, so it is easy to use from the very first call.

How to Use the Demo:

Perform any of the following transaction types supported in the demo.

- Check Account Balance
  - —"What's my savings account balance? (or checking or VISA accounts)?"
  - —"Check my checking account balance."
- Review Recent
  - —checks
  - —withdrawals
  - —deposits
  - —transfers
  - —bill payments
  - —transactions
  - —press * to interrupt the list and return to the menu
- Transfer Money
  - —"Transfer money from savings to checking" (say HELP to hear a full list).
  - —"Transfer $325 to checking from savings."
- Pay Bills
  - —the phone company (or Pacific Bell)
  - —Bank of America (or BofA, or the mortgage)
  - —VISA
  - —American Express (or Amex)
  - —New York Times (or the paper, or the Times)
  - —Pacific Gas and Electric (or the power company, or PG&E
  - —say HELP to hear a full list of payees

At any point one can say "HELP" to get more information. Just say "goodbye" or hang up to end the demo.

ANOTHER HELP DESK PACKAGE

Another example of CTI-enabled Help Desk software is Advisor from Decisif, a small Canadian software development company. It provides the following features:

- generation of problem identification tickets
- tracks support calls from inception to problem resolution
- Escalation of calls based on time elapsed or nature of call
- Built-in knowledge base and inventory module
- Report generator provides realtime information on activity levels, resource constraints, and agent performance
- Call information can be access over the Web interface and agents can update, close, and transfer calls off site

Decisif describes its CTI products as completely "open." For example, end users can add additional applications from the base package purchased from Decisif by programming with:

- ActiveX
- OCX (Ole custom control from Microsoft; allows prewritten packages of code to be downloaded from Web servers)
- OLE (object linking and embedding; allows windows programs to exchange information)
- DDE (dynamic data exchange; allows two programs running under Windows to share data)
- TSAPI
- TAPI

The company also supports database access via:

- DBF (dBase, FoxPro)
- ODBC (Open Database Connectivity; allows various databases to be accessed via a common interface)
- MDB (MS Access)
- SQL (SQL server)

Decisif is only one of hundreds of vendors that have elected to make their CT packages open to changes. In evaluating platforms/packages, the vendor's tools and ability to interface should be examined carefully to ensure that packages can be linked and modified as needed for the organization's needs. Telephony applications need to be modified as much as traditional IT systems.

SUMMARY

CTI technology — from the simplest Windows 98 dialer to sophisticated routing/database lookup systems to advanced speech recognition — is becoming a standard tool of the communication industry's repertoire. As telephony functions become easier to add to existing applications via middleware and standards-based development products, users will begin to appreciate and demand even more telephony functions in their day-to-day business applications. The wise organization will develop an infrastructure and conceptual architecture *before* users begin demanding product on the desktop.

Chapter 5
Unified Messaging

OFFICE WORKERS MUST CHECK MANY LOCATIONS FOR THEIR MESSAGES —
e-mail inbox, fax machine, office voice mail, cellular voice mail, alphanumeric pager, paper-based mail, and assorted display devices (e.g., Amtel
for quick display messages, display boards, overhead paging). Unified messaging (UM) attempts to reduce the user's burden by combining some of
these technologies, at least from the user's perspective, into a single presentation. Using a common GUI (graphical user interface), users can logically save, retrieve, forward, and convert messages from one format to
another. In today's "mongrel" world of many message technologies, unified
messaging has the potential to shorten the learning curve and simplify the
user's communications world.

An ancillary benefit of unified messaging is the ability to transcend some
of the limitations of the century-old voice network and move, at least partially, to more bandwidth-efficient, IP-based transport architectures. One of
the challenges of UM is to avoid introducing so many fundamental architectural changes in the communications environment that it never gets implemented. Interdisciplinary coordination and intense planning are required
by representatives from LAN and WAN management as well as telephony,
and human resources (the latter to feed directory services).

BENEFITS OF UNIFIED MESSAGING

Given the considerable effort required to implement large-scale UM
within an organization, a listing of potential benefits is a good place to
start. Support from the end user and multiple departments is needed in order to implement without system failure, clogged LANs, etc.

Following are features found in most upper-end unified messaging packages.

- single-user interface (GUI) for voice mail, fax, e-mail, and potentially
  video clips
- access to various voice mail features via point-and-click
- ability to save and view messages in folders
- ability to view header information (important for laptop users) with
  priority indicator
- archive option for all messages, regardless of media (most packages
  have proprietary compression that is superior to standard .WAV files)

87

- forward, reply, or call sender from desktop or standard telephone
- create mailing lists
- visual or audio notification of new messages
- phone book (preferably integrated into an LDAP-compliant directory)
- ability to change coverage (i.e., where to send calls when called party not present)
- voice annotated fax messages
- privacy of received fax messages
- outgoing folder support (showing message status, delivery reports, and message resubmit options)
- print new message folder information (for a callback list); also, ability to capture the number and redial by point-and-click
- print message summaries
- schedule delivery of outgoing messages
- transfer fax image to other desktop/Windows applications
- nonsequential view/listen to messages
- add header/subject annotations for filters, comments, priority levels
- option to listen or record messages via sound card/microphone or standard telephone set

Other miscellaneous benefits include:

- avoid physical printing of many faxes
- more efficient use of analog ports (fewer physical fax machines needed; more efficient use of ports)
- physical fax to physical fax now replaced by UM to UM on desktop
- UM can be linked to CTI functions to further enhance user efficiency (for example, conferencing, call hold, call drop, redial, screen pops, making calls from the voice mail directory, and other functions can be linked into the UM software — provided the CTI server and software has been implemented)

BASIC ARCHITECTURE

Exhibit 5-1 shows two options for UM. The first option gives users a choice of listening and recording their messages via sound card and microphone or via the telephone. The second option provides the same services except that sound is solely via the telephone.

Sample desktop user interfaces are shown in Exhibits 5-2 and 5-3. The reality of unified messaging is more complex than these illustrations would suggest. First, in organizations that already have an installed base of applications, there is competition for space (memory, etc.) on the desktop. Lotus Notes users, for example, may prefer to have unified messaging as a background function and the interface would remain the standard Notes interface. All that would change is that new media types would appear — voice and fax attachments along with traditional e-mail messages. The sec-

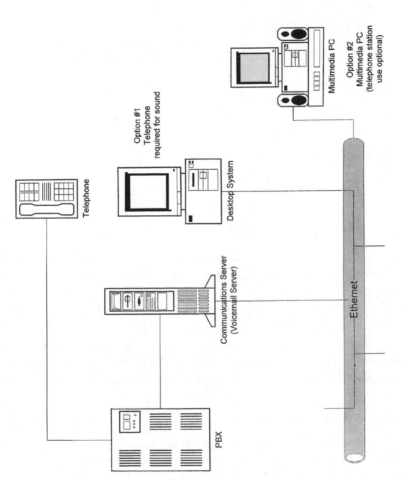

Exhibit 5-1. Basic unified messaging architecture.

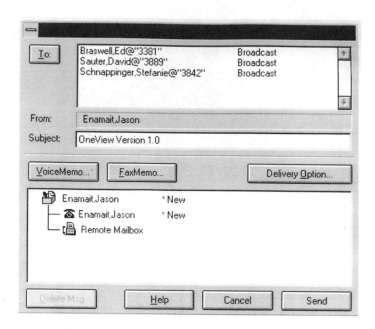

Exhibit 5-2. Unified messaging screen (Courtesy Centigram).

ond complication is deciding where the messages will reside. Will e-mail, fax, and voice mail all reside in the voice mail server (e.g., the Lucent Intuity Audix)? Or will the "original" copy of e-mail reside in the Notes server and copies be made for the unified messaging server? If so, where does the message/attachment get deleted?

One of the risks of UM is that if the system goes down, e-mail, fax, and voice mail are all down at the same time. For that reason, Nortel has elected to use a distributed architecture, where separate servers are used for each type of media.

Unified messaging uses a number of CT-intensive functions, including:

- text-to-speech to read e-mail over the telephone
- OCR to read either headers or the entire text of incoming faxes
- text-to-fax (if that is all the recipient has available)
- speech recognition (just now developing for UM)
- conversion to paging alphanumeric text (limited)
- directory services

Some of the functions listed above are driven by the obvious fact that users do not always sit in front of their PC or terminal. Travelers, employees in sales, and employees who take work home at night can use these functions. For example, Lucent's Octel Unified Messenger allows users to listen

Exhibit 5-3. Unified messaging screen (Courtesy Nortel).

to e-mail via text-to-speech, then respond with voice messages. Using Microsoft Exchange as a technology base, users can forward fax, e-mail, or voice messages with voice comments to others on their corporate e-mail network or over the Internet.

VARIETIES OF UM AND DESIGN CONSIDERATIONS

Articles and vendor literature on UM use a number of differing terms that are just similar enough to confuse the public. The United Messaging Consortium (www.unified-msg.com) offers the following definitions.

- Unified mail: the ability to store messages of all media type — voice, fax, e-mail, video, paging, etc. — in one mailbox with accessibility from either a PC or telephone.
- Unified Messaging: the ability to create and respond to multimedia messages with fidelity to the originator from either a telephone or PC (especially across different vendor platforms); additionally, personal call control permits real-time control of incoming calls and call rebound with message processing.
- Universal messaging: the ability to create any type of message and to send it to anyone without regard to the recipient's mailbox requirements.

Each one is a superset of the previous definition. Additionally, the following terms are used and can be confusing.

- Integrated mail: an implementation of unified mail; combines several message stores to present a single view to the end user.
- Integrated messaging: an implementation of unified messaging; combines several message stores to present a single view to the end user.

Unified messaging has focused on incoming messages. However, from a long-term perspective, messages must be sent out to various end points that may have only a telephone or perhaps incompatible equipment. If the originator of a call has only a telephone and sends a message that winds up in a unified messaging end box, how does the "reply" work? If the originator has voice mail, is it AMIS (audio message interchange standard) or VPIM (Voice Profile for Internet Mail) compliant?

Under UM, distribution lists must be examined. How is the system to know the capabilities of the recipients? Should the lists include only those with similarly equipped mailboxes?

Unified messaging will become far more widespread when disparate systems can communicate and retain messages. Faxing is a good analogy. Before the days of Group III standards, faxing languished because of incompatible protocols. As soon as the industry wholeheartedly adopted Group III, faxing use went up exponentially. Telephones, using the PSTN, connect regardless of vendor. The caller is not concerned with architecture. UM, on the other hand, requires the user to be conscious of sound cards, versions of e-mail and attachments, extension compatibility, etc.

Ultimately, UM is more about getting to messages in one logical place from any physical place. This definition solves the problems of travelers, people without laptops or PCs, and others. the goal of UM should be to hide the technology, at least to the extent possible.

VOICE PROFILE FOR INTERNET MAIL (VPIM)

AMIS has, up until recently, been the only means by which unlike voice mail systems could send messages to each other. Although AMIS works, it has limited functionality. It will, for example, allow messages to be sent from Lucent Intuity Audix (via Lucent Interchange) to a Rolm Phonemail system. But return receipt is not available, nor is prioritization.

The new paradigm for Internet messaging is VPIM. The goal of VPIM is to establish an internationally accepted standard that will allow the transparent interchange of both voice and fax messages between disparate voice messaging systems as well as non-voice e-mail systems (MIME compatible). In addition, and critically important, is to establish a directory service to look up routable addresses. If a phone mail message is to be sent to an e-mail inbox, there must be a link between a PSTN phone number and an Internet ID.

Although VPIM is an emerging standard, within two to three years it will very likely be accepted by most major voice mail and e-mail vendors. SMTP (Simple Mail Transfer Protocol) and MIME (Multipurpose Internet Mail Extensions) are becoming the common denominators for e-mail. VPIM carries this further by specifying how voice mail servers will exchange messages — including UM servers. By limiting the standard to server-to-server and ignoring the client side, VPIM is considerably easier to implement in existing infrastructures. State-of-the-art users can point, click, and talk into a microphone to send messages, while traditional telephone-only-based users can hear the message on existing equipment. Of course, VPIM "helper" applications to play back messages will be needed on the desktop.

Bernard M. Elliot, a consultant from Vanguard Communications Corporation, notes that many users want to "intentionally message." In other words, they specifically do not want to call the party, let the phone ring, then wait for voice mail to engage. They want to quickly compose a message and send it. With proprietary voice mail systems, users can do this internally, but these features are not generally available to outside parties. Intentional messaging is an extension of distribution lists, forwarding, and reply functions that are so commonplace now in workgroups.

UM and voice mail systems using VPIM will need to use an X.500-related directory service such as LDAP (Lightweight Directory Access Protocol). Because LDAP can be installed on many legacy systems, it should provide for further expansion of UM capabilities outside a single network. Application servers can read or update an organization's home-grown directory using LDAP. Then users can enter a phone number and the directory will do a lookup on the address and possibly hear a name spoken back to them, confirming they have the right party.

While VPIM has a long way to go, it is being embraced by many of the larger voice mail vendors, such as Centigram, Nortel, Siemens Business Communications, and Lucent/Octel. Also, e-mail vendors that support IMAP4 and the older POP3 will certainly support VPIM.

A UM PACKAGE CHECKLIST

There are dozens of vendors in the UM market. Many offer only partial functionality or do not integrate well with existing infrastructures. Following is a checklist of functions that should be considered when reviewing specifications.

- access virtually all functions with laptop or telephone
- unified (single input) administration

- support for industry standard voice formats (VOX, WAV, PCM, and AD-PCM), so that others outside the organization can read forwarded messages
- integrate and use existing name and address book (e.g., Lotus Name and Address book)
- operate with multiple operating systems on the client side (e.g., UNIX, Windows 98, NT, Linux)
- advanced searching, sorting, and macro capability at the desktop (for example, be able to review new messages, messages from a specific time period, messages from a specific person or with a text string in the subject line)
- read fax over the telephone via OCR
- logical and robust means to keep messages in sync — how do parallel links maintain their integrity (i.e., message deleted on one server but not another)?
- browser access to messages
- select order of listening to messages
- fast forward, rewind, and pause messages
- ability to have two-way recording
- record specific greetings for holidays, time of day, etc.
- audiotext and bulletin board (for simple messages not requiring IVR or CTI implementations)
- remote maintenance while the system is operating
- administrator-selected password length and type
- call transfer
- dial by name or extension
- fax detection routing and notification
- off-site mailbox access
- prompt/greeting override
- configurable number of retries on no DTMF
- realtime display of system activity
- touch-tone telephone options
- unlimited or selected recording time length
- broadcast with and broadcast without message waiting light notification
- append new message to old message
- busy versus no answer greetings
- follow me (call forwarding)
- message count
- message delete and undelete
- message receipt confirmation
- outbound message notification
- purge time specification
- time and date stamp
- urgent message delivery or paging
- volume controls

SUMMARY

Unified messaging can and will enhance productivity by simplifying storage, transmittal, and access to messages of different media via PC, telephone, fax, pager, cellular phone, and other devices to come. The promise of UM is that the mechanics (form) of the message will move to the background while the content will move forward, being more accessible than ever.

While it will require considerable effort on the industry's part to make this interconnection work, when complete, it will be *de rigeur* for any organization with significant communication needs. Vendors that do not currently support underlying standards (such as VPIM) should be regarded with suspicion — they must necessarily play heavy catch-up in a few years.

Chapter 6
Call Center Management

THE CALL CENTER HAS BEEN THE TRADITIONAL DRIVER OF COMPUTER TELE-PHONY INTEGRATION as well as many other telephony technologies. The downsizing of call centers, enabled by the availability of new communications technologies, and the exponentially increasing power of chips and DSP boards have helped spread call center technology to a much broader market.

An advanced call center is a virtual arsenal of technology. Mature call centers typically have fully developed reporting systems so that even slight advances resulting from automation can be validated and used to justify new applications. The following are some of the important technologies found to today's call centers:

- ACD
- CTI
- interactive voice processing
- workflow
- predictive dialers
- image processing
- e-mail processing
- video/multimedia interaction
- audio text and announcers
- fax on demand
- management reporting, including workforce management tools

Many of these tools provide dramatic, objectively measurable benefits to the organization. IVR alone can reduce approximately 40 percent of calls that would normally go to a live operator. Appendix D lists several success stories that illustrate the business value of these telephony tools.

The following chapter sections describe some of the more important technologies and strategies found in medium to large call centers.

PREDICTIVE DIALING SYSTEMS

The outbound group within a call center needs to talk to as many live prospects as possible. Without an automated predictive dialer, an agent, on average, will contact only 30 to 35 people per 100 attempts, wasting much of the agent's time. To present a live human to the agent, predictive dialers go through the following scenario.

- Determine if any outbound agent is available.
- Search through a database of potential customer numbers.
- Launch as many calls as possible.
- Filter out those that are answered by answering machines, fax machines, modems, etc. Most predictive dialers look for a long string of words to indicate an answering machine ("Hello, you have reached the Smith's residence...") versus the human response of "Hello" and then silence. Some of the more sophisticated systems measure the electronic voltage and line noise produced by an answering machine.
- Present to the free agent (with 70 to 90 percent reliability) a live prospect. Most of the time, the agent hears no ringing or delay, but just hears the familiar "zip" of the headset, indicating that the live prospect is on the line. The computer does most of the routine work. This has a positive effect on call center staff, who can become fatigued with the constant routine of finding the number, typing it in, listening to ringing, etc.

The software for a predictive dialer is complex. It must consider (using mathematical models) the trunking capacity of the call center, the number of agents, the time between calls needed for maximum agent efficiency, the length of an average conversation, wrap-up time, and other factors. It is not a weekend project with Visual Basic and TAPI 3.0.

A robust dialer should be able to obtain sales candidates from various commercially available lists based on geographic, demographic, or other characteristics. Federal and some state statutes require that a "do not mail to or call" database be maintained. Finally, the telephone number selector should have the capability of quickly changing parameters. For example, if the sales organization is promoting a product that appeals to lower-income households, the predictive dialer should be able to easily provide a list based on a higher level of household income when the marketing direction changes to upscale products.

Some other desirable features include:

- manual entry of the phone number
- capacity to specify multiple sales campaigns simultaneously
- recording of DTMF (handset keypad) tones or voice answers; if caller is interested, call can be presented to the live agent
- remote administration for administrators
- logging features and summary reports
- open architecture (e.g., database is compatible with MS Access or Visual FoxPro)
- survey capability

Articles and books often give the impression that predictive dialers are solely for marketers who are "pushing" products (rather than a Web site that "pulls" customers). Reality is somewhat different: predictive dialers are used for a diverse set of needs. Examples include (1) charitable organizations that use the technology to call homebound senior citizens to check the status of their health; (2) mortgage investment companies that broadcast a special or time-sensitive lending rate to a selected population; and (3) a large number of people who can be notified by public health or safety officials that an emergency condition exists.

Predictive dialing is not for every outbound call center. It functions best for short calls that require little time for the agent to review preliminary materials. Calls to VIP customers should not be made without an investment in agent time and consultation with the customer's history.

Predictive dialers run on a variety of hardware. Low-end systems need only a PC running NT or Windows and a telephony card. The cost varies from around $1000 to $10,000 for high-end systems.

CALL CENTER MANAGEMENT AND STANDARDS FOR AGENT PERFORMANCE

Unlike many business functions where management decisions must be made without the benefit of strong numerical support, the call center is (or can be) awash in statistics, trends, graphs, and exception reports. Reports and measurements, if well-designed, drive agent behavior (organizations get what is measured and rewarded). According to TCS Management Group (www.tcsmgmt.com), 60 percent of the average call center's budget is spent on personnel. Of the remainder, 25 percent is for networking and long-distance charges, 5 percent is for equipment, and 10 percent is for overhead.

Some of the business-level questions that should be asked of the organization's call center management reporting system include:

- Does it accurately reflect agent performance?
- Does it take into account all the technologies used (e-mail, ACD, IVR, etc.)?
- Does it address caller satisfaction?
- Does it identify opportunities to improve the speed, quality, and turnaround of caller problems, questions, or order fulfillment?
- Does it take into account external metrics as a basis of comparison?

The workload of a call center arrives randomly. Overall volumes fluctuate considerably over time, making staffing planning difficult. Nonetheless, patterns emerge over time that enable proactive scheduling and other peak load techniques to handle variable loads.

Call center metrics are shown in Exhibit 6-1. These are some of the major metrics used. Well-established call centers will have these and others more appropriate to their business. Exhibits 6-2, 6-3, and 6-4 show a call volume forecast, a net staffing variance report, and a daily work report, respectively.

Some of the features to look for in call management software include:

- forecasting based on mathematical models such as moving average, Box-Jenkins, and simulation techniques
- calls received by call type (see Exhibit 6-5).
- calls by time of day
- total sales by month, quarter, year
- scheduling for an arbitrary number of days into the future
- agent productivity, realtime
- call blending (inbound and outbound) forecasting
- Web-enabled reporting
- special reports on meetings, breaks, and vacation times
- agent performance
- agent share of workload
- trunking reports (idle trunks, percent blockage)
- longest call waiting by gate (in call center terminology, a "gate" is a trunk group of telephone call groupings that is set up such that any agent receiving calls from the same gate can answer any of those calls)
- gate traffic
- export capability into programs such as Microsoft Access, Crystal reports, or other programs that use ASCII delimited file formats
- multi-site reports that combine call center reporting into an aggregate, enterprisewide portfolio of information
- agent time by category for all agents or for those in a specific group/gate (see Exhibit 6-6)

It is important to keep call center management informed, as well as the agents themselves. Call center terminals (telephones) should have an easy-to-read LCD display that can show realtime statistics. For example, high-end agent display sets from BCS Technologies (www.bcstechnologies.com) display the following information:

- calls waiting
- calls answered overflow
- longest call waiting
- calls diverted
- average speed of answer
- service level
- lost calls

Exhibit 6-1. Call Center Metrics.

Metric	Explanation	Suggested Values
Mean abandonment time	How long a caller waited before hanging up	25–65 seconds
Mean handle time	Total talk time and administrative work time to complete the transaction	10 minutes for general call center activities; 15 minutes for technical support
Mean hold time	Seconds that an agent puts a customer on hold	25–70 seconds
Mean answer speed	The total time a caller is in the queue, divided by total calls answered	< 22 seconds
Mean number of rings	How many rings the customer must listen to before the agent or IVR answers the call	2–3 rings
Mean talk time	Average number of seconds each agent spent with a caller	200–400 seconds for most organizations; < 9 minutes for technical support
Percentage of one-time-only calls	The percentage of calls that do not require a subsequent call or follow-up	> 82 percent
Mean queue time	Average number of seconds a caller is on hold waiting for an agent to answer; this information comes from ACD reporting	25–100 seconds
Agent attendance ratio	Over a reasonable period of time (e.g., a month or quarter), divide the number of shifts worked by the number of shifts that the agent was scheduled to work	> 93 percent
Calls per standard workday	Number of calls in a standardized workday (e.g., eight hours); prorated for longer or shorter days to ensure consistency	400–800 calls for simple inquiries and transactions; 40–60 calls for highly technical/complex transactions
Abandon rate	Defined as a percentage of calls abandoned relative to total calls received. An abandoned call is one received by the call center but dropped by the caller before reaching an agent or information source (such as an announcement or complex IVR)	< 6 percent
Agent busy percentage	Equals talk time plus time caller is on hold, divided by time agent is on duty	> 88 percent
Agent turnover	Number of agents who left the organization in a quarter, divided by the number of agents present at the beginning of the period	< 20 percent
Percent of trunking blockage	Percent of inbound calls blocked because of insufficient trunk capacity	Erlanger value of P.01 or better (no more than 1 percent blockage)
Percent of calls transferred to another agent or IVR	Percentage of an agent's total calls transferred to another agent or automated system (such as an IVR)	< 4 percent
Service objective percentage	Calls answered in less than X seconds, divided by total number of calls	85 percent of the calls answered in less than 20 seconds
Total calls coming into the call center	Includes every possible measure such as calls processed, blocked, and abandoned; used for overall capacity planning	Varies by size of call center and business

Exhibit 6-2. Sample report: daily forecast of call volumes.

Date	Weekday	Daily Total	Weekly Total
1/31/00	Monday	8,728	
2/1/00	Tuesday	6,331	
2/2/00	Wednesday	7,374	
2/3/00	Thursday	7,869	
2/4/00	Friday	7,840	
2/5/00	Saturday	8,665	
2/6/00	Sunday	6,220	46,807
2/7/00	Monday	7,816	
2/8/00	Tuesday	7,918	
2/9/00	Wednesday	6,377	
2/10/00	Thursday	7,810	
2/11/00	Friday	7,884	
2/12/00	Saturday	7,673	
2/13/00	Sunday	8,368	53,846
2/14/00	Monday	6,036	
2/15/00	Tuesday	8,337	
2/16/00	Wednesday	8,748	
2/17/00	Thursday	7,092	
2/18/00	Friday	8,466	
2/19/00	Saturday	8,549	
2/20/00	Sunday	6,641	53,869
2/21/00	Monday	6,012	
2/22/00	Tuesday	7,292	
2/23/00	Wednesday	7,887	
2/24/00	Thursday	7,967	
2/25/00	Friday	6,096	
2/26/00	Saturday	8,985	
2/27/00	Sunday	8,214	52,453

- agents manned
- calls offered
- agents busy
- calls answered
- agents available
- calls answered primary
- agents unavailable

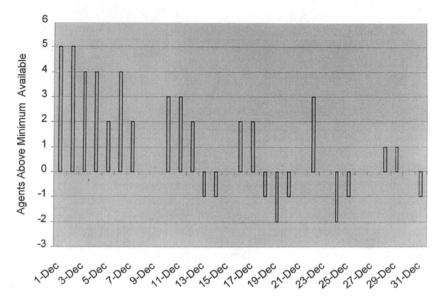

Exhibit 6-3. Net staffing variance graph.

Exhibit 6-4. Daily work report.

Employee Name	Start Time	End Time	Off Duty 1	Off Duty 2	Off Duty 3	Work Type	Supervisor
Einstein, Wilhelmina	500	1400	550	750	900	Follow-up	Rommel, Erwin
Francisco, Mary	530	1430	900	1100	930	Sales #1	Rommel, Erwin
Grant, Butch	1000	1900	1100	1300	1400	Sales #3	Rommel, Erwin
Hill, Al	830	1730	1000	1200	1230	Sales #2	Stansbury, Ginny
Lee, Bob	830	1730	1100	1300	1230	Sales #4	Rolundo, Erwin
Longstreet, William	1100	2000	1300	1500	1500	Sales #2	Ruf, Patricia
Mosby, Bubba	700	1400	815	1015	1100	Sales #1	Ruf, Patricia
Rex, Libby	600	1500	930	1130	1215	Follow-up	Ruf, Patricia
Smith, Geraldine	600	1500	1130	1330	1215	Sales #4	Ruf, Patricia

---

Large, wall-mounted display boards, such as those from Inova Corporation (www.inovacorp.com), ensure that all agents and supervisors are fully aware of call center statistics realtime. Inova's LightLink Server, for example, displays a variety of service-level statistics on large electronic boards and also sends relevant information out via pager, e-mail, fax, intranet, desktop monitoring software, cellular telephones, and video monitors. Providing information to all call center employees promotes self-management and helps them to focus on critical activities.

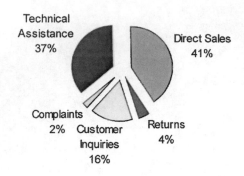

Exhibit 6-5. Calls received by category.

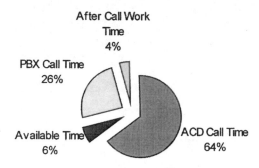

Exhibit 6-6. Agent time by category.

Some features to look for in a wall-mounted display include:

- ability to accept call center statistics from a variety of sources
- large, multi-color LEDs (red, green, and yellow preferred)
- audible alert tones for new messages
- accurate time
- multiple fonts
- animation

The more advanced packages allow flexible selection criteria so that supervisors can target specific calls or agent groups. For example, reports could be generated by shift, agent group, campaign type, or for a specific agent. In addition, customer-centric reports based on account ranges, calls from specific regions or area codes, and VIP versus first-time customers can be created.

One example of an agent scoring system is Eyretel's QualityCall agent evaluation system (www.eyretel.com). The package scores agents in categories such as:

- product knowledge
- sales closure
- selling skills
- customer experience
- control of the call

Each category above has subcategories. For example, sales closure is further defined by ratings such as "asks for the sale" and "recaps all agreements/savings." The supervisor or manager can target the agents or campaigns that should be recorded. Later, the selected group can be analyzed in depth to determine, for example, if agents have been adequately trained in the products or services offered. Exploratory analysis can provide insight into problem areas. For example, analysis of agent performance at the beginning of a shift versus performance at the end of a shift may indicate a scheduling problem.

Analysis categories include:

- agent identity
- calling type
- campaign number
- caller ID
- dialed number (DNIS)
- agent group
- DID (direct inward dial)
- customer account number(s)
- claims number (for insurance companies, etc.)
- customer type

The system keeps benchmark scores (as a standard) to provide a basis for evaluation of current scores.

GOOD IVR DESIGN

Interactive voice response systems are key to reducing the quantity of questions and information that must be furnished by agents. However, poorly designed IVRs will result in caller frustration and a large number of "zero out" selections to a live operator. The following are some design criteria for effective IVR systems.

- Standardize on yes or no response. Most users expect 1 for yes and 2 for no.
- Use # at the end of user input — particularly when the input is variable in length.
- Do not require the user to input multiple digits (such as 33). If more than ten selections are needed, a second level of options should be used.
- Do not commingle various voices. It is disconcerting to the user to hear, for example, a male voice say, "Press 9 for more options" and

then hear a female voice say, "Press # when finished." Disparate voices destroy the professional image of the IVR application.

- State the action first, then the number. For example, say, "For sales, press 1" rather than, "Press 1 for sales."
- Provide opportunities for callers (especially repeat callers) to bypass an announcement or menu when they know what they want.
- Simplify scripts. Designers sometimes appear to have the cartoon character Elmer Fudd in mind when they design scripts. Customers are quicker to comprehend than some IVR developers believe.
- Avoid the prompt, "Do you have touchtone?" or at least put it last in the initial script. Most callers now have touchtone. Those that do not should be told to stay on the line for an agent. Another alternative is to implement speaker-independent voice recognition for the digits.
- Avoid use of alphabetic characters for input. It may be "neat" to have the caller spell out the name "lawn mower" in an IVR shopping application, but the caller will not appreciate the elegance. Also, Q and Z are missing on a standard telephone.
- Take advantage of ANI and DNIS. If the customer can be identified by calling number, he or she can be taken to menus more likely to be of interest.
- Send callers to the appropriate menu based on language if a significant percentage of the customer base speaks more than one language.
- Include systems to analyze and report IVR efficiency, including the number of times the caller zeros out to a live agent. Use surveys if appropriate.

AGENT RECORDING AND MONITORING

Agent recording is used to monitor the quality of agent interaction with the customer, to confirm orders, and to review the efficiency of the overall call center design. Software tools must be in place to go back to a recording based on time of day and date.

Recording may be full-time, selective (sampling), or event driven, as when a caller presses a specified key in an IVR application indicating he is placing an order. The call center should always have the ability to record on demand, based on actions initiated by the supervisor or agent.

In order to retrieve the call, information such as time, date, agent, ANI, DNIS, call count, marketing campaign, and any available IVR recorded data should be attached to the recorded voice conversation. There are a number of media that are appropriate for recording, including hard drive for quick retrieval, a near-online jukebox for older recordings, and a DAT (digital audio tape) for archived records. Typically, retrieval can be by telephone playback or PCs with speakers, export of .WAV or .AU files via the LAN, or remote access.

Exhibit 6-7 shows a recording architecture in a call center.

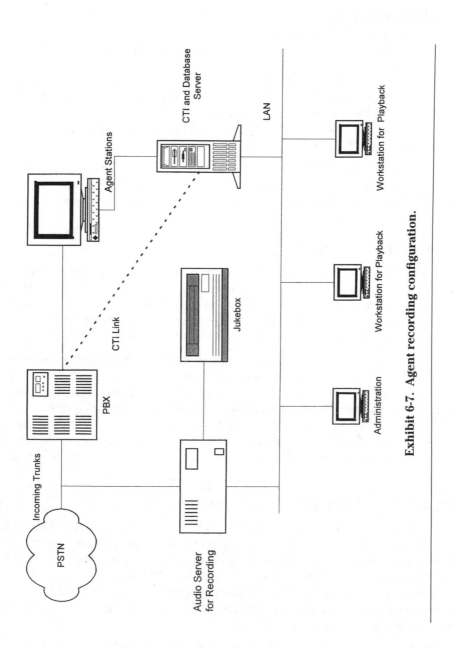

Exhibit 6-7. Agent recording configuration.

Some basic questions to ask when reviewing recording alternatives include:

- Does the system have adequate capacity to handle volumes even during peak periods (such as holidays)?
- Is the indexing adequate to ensure quick access to previously recorded calls?
- Can the system be integrated into a multimedia environment so that data, audio, and potentially video can be recorded as a unified entity?
- Can calls be accessed over the LAN?
- Can recording be turned on and off easily, and can it support selective recording?

In some call centers, agents do not sit at the same desks all the time or job-share so that different agents sit at the same terminal. In these cases, it is important that the recording be tied into the login ID of the agent and not the extension or address of the physical terminal.

INTEGRATED FEATURES

Call center technologies should be integrated. For example, many call centers offer a series of announcements to callers — nothing more than spoken information with no options for the caller. A well-designed ACD queue will allow callers on hold to exit to an announcement, listen to it, and then return to the queue without losing their position.

Two other integrated features enhance the call center's effectiveness: agent whisper and automated callback. Agent whisper presents key information to the agent regarding the call that is just about to be received. This helps the agent to know the expectations of the caller (e.g., whether they have called in on a complaint line or a merchandise ordering line). Automated callback gives callers the option to leave their phone number and record their name. They can then be called back during non-peak hours.

WEB INTEGRATION AND THE MULTIMEDIA CALL CENTER

Affluent households are more likely to shop or seek information from the World Wide Web. Hence, full-featured call centers increasingly have Web agent software. Direct interaction with customers over the Web can be implemented in a variety of formats: text chat, IP telephony, and callback requests. In addition, all the traditional reporting, routing, and management capabilities found in traditional call centers are increasingly embedded in Web agent software.

To generalize further, an integrated call center includes e-mail, fax, Web communications, and video, as well as traditional voice communications. Integration maximizes service by allowing customers to use the media they prefer. It also ensures that the customer's complete history and transactions are available for comprehensive services.

The multimedia call center enables self-service (24-hour availability) with the option of having a live representative. The use of voice recognition further enhances the self-service concept by allowing the user to quickly ask for information, pricing, etc., directly, rather than navigating through complex and time-consuming IVR menus. By providing considerable automated information to the customer up front, agents' time can be spent on things that are of higher value. For example, some customers prefer to give credit card numbers to a live agent; others want instructions.

Some of the means by which customers interact with the call center include:

- live interaction via telephone
- self-service via IVR
- leave voice mail
- Web site for self-service
- auto-acknowledgment of e-mail (such as RapidReply's auto response e-mail gateway, www.rapidreply.com). This feature allows the call center to return diverse information such as catalogs, order forms, coupons, price lists, real estate agents, etc.
- personalized response via e-mail
- voice over the Internet (either with two lines or via the same line)
- Internet chat
- paper (postal) mail; can be scanned for later use by sales representative
- receipt via fax
- auto-response via fax; use OCR and word spotting on incoming fax to generate appropriate response

The integrated call center will allow more personalized services. On the Web, cookies (personal tokens), registration IDs, and the open profiling standard (proposed by Netscape, Firefly, and Verisign, Inc.) allow the call center to identify a returning customer. In the same way that ANI can be linked to a database to display screen pops profiling a customer, Web software can identify a customer's habits to the agent. As Internet technology improves, it will be easier to deploy work to home agents, who will have access via the Web to voice, video, and data collaboration.

In order to integrate media beyond voice, call centers must rethink their processes and systems. Areas where design changes may be required include:

- Customer database and profiling.
- Documentation and design of the flow of calls. Flowcharts should be developed that show the flow of customer communications as they come in from various media: direct from an ACD, via a IVR, e-mail, fax,

video from a kiosk, etc. The flowchart will show actions to be taken based on customer choices and will tag appropriate service levels at those branches.

- Staff capabilities. Skills need to be further defined.
- Call center organization and scheduling. Web interaction, for example, may be more evenly distributed over a 24-hour day than voice communications. International Web browsers will call during off-hours.

EXAMPLE INTERNET CALL CENTER

To illustrate Internet call center technology, selected features of Lucent Technology's CentreVu Internet Solutions are described below (adapted from a Lucent Technology's white paper on Internet solutions, www.lucent.com).

Basic features include:

- Escorted browsing. Agent can "lead" customer to the Web page(s) necessary to bring the transaction to a close quickly, as the customer browser will synch to the page(s) the agent sees.
- Multimedia call transfer/conference. Improves responsiveness to customer needs for more information by allowing agents to transfer or conference complete Web/voice sessions.
- Multimedia queuing. Enhances queue experiences by allowing customers to continue browsing while in queue or providing infomercials that enlighten or entertain customers.
- Single-line voice call/Web browsing. Customer satisfaction from uninterrupted Web browsing sessions — no callback, no second line needed. Voice calls go straight through to the call center agent.
- Universal agent. Optimizes resources and saves money by using existing agents to handle both Web/voice and voice-only calls.
- Web pop. Improves response time to customer requests, as the agent screen shows the page the customer called from.
- Web voice calls. Provides convenience to the customer; reduces incoming call expense to the business.
- Web/call center integrated call management reports. Allows businesses to evaluate the sales effectiveness of promotions, campaigns, etc., and the level of service provided to customers, for each voice-enabled Web page.

For these technologies to work, the customer needs a PC equipped with a Java-enabled Web browser such as Netscape Navigator or Microsoft Internet Explorer. To use the Internet for voice calling, the PC must be multimedia equipped, with an H.323-compliant telephony application such as Microsoft NetMeeting 2.0 or later, and a reasonably high-speed connection to the Internet (minimum 28.8Kbps). The H.323-compliant telephony applications operate under the latest Windows NT operating systems.

In the call center itself, Lucent requires their Internet telephony gateway, Definity PBX (they refer to the PBX as an Enterprise Communications Server), and a POP-3-compliant mail server such as their Intuity Audix system with Internet messaging.

In addition, high-speed LAN connectivity and Adjunct Switch Application Interface (ASAI) software provide the links for the Definity PBX, CentreVu Computer Telephony server, ITG, mail server, and agent PCs. Expert Agent Selection (EAS) software works with Definity Automatic Call Distribution (ACD) to queue and route Internet calls, e-mail, and fax the same way that ordinary calls (e.g., incoming "800" calls) are routed within the call center

The call center agent desktop must be similarly equipped with a PC and the appropriate 32-bit operating system — for example, Windows NT — and a Java-enabled Web browser. No desktop speakers, microphone, or telephony application is needed because the voice portion of the call is delivered through the Definity PBX to a compatible telephone, such as a Callmaster voice terminal.

In a scenario where the end user is equipped with a multimedia-enabled PC, the customer uses a Java-enabled browser to access the World Wide Web. While browsing a Web site, the customer simply clicks a button on the Web page to talk to a customer representative. There is no need for the customer to end the browsing session and wait for a callback. The customer simply uses the Web page interface to request the type of call he or she wants — an Internet voice call, a text chat, or even a callback on an ordinary telephone using the Public Switched Telephone Network (PSTN) (see Exhibit 6-8).

The application initiates a call across the Internet to the call center ITG, utilizing the same telephone line used to connect to the ISP. The ITG downloads a call control applet, written in Java, to the customer's PC to start the Internet telephony application. (An applet is a "mini" application program that is created in the Java language and run on a PC using a Java-enabled browser.) The call control applet provides the customer with call status messages as well as the interface through which the customer drops the call, conducts text chat, or types in data to collaborate with a call center agent (see Exhibit 6-9).

Exhibit 6-8. Web customer contact screen.

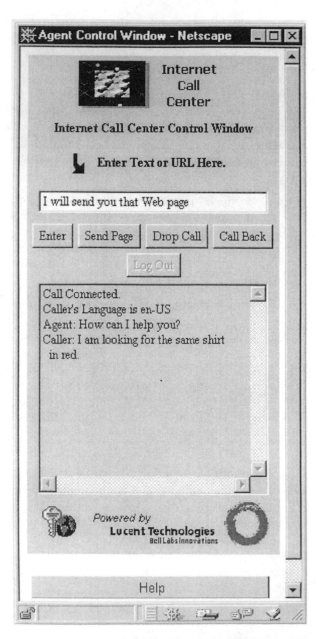

Exhibit 6-9. Call control applet screen viewed by customer.

The electronic messaging portion of the package allows organization's to manage:

- e-mail generated directly from the Web site (e.g., customer responses to "Write to us now" links created for specific Web pages and routed to skilled agents)
- e-mail sent to a specific address (such as direct to the call center) and routed based on a corresponding VDN
- faxes forwarded to an electronic mailbox and stored in a POP-3-compliant mail server

In a typical scenario, a customer sends an e-mail to the service center, either through a typical e-mail channel or by filling out a form on the Web site (accessed, for example, through a "Write to us" link). The message travels over the Internet to the premises mail server. A "phantom call" through the Definity PBX — creating a virtual message call that is treated like an ordinary voice call — is queued and routed according to the VDN and skills set specified by the application.

When an agent becomes available, the e-mail "call" is delivered to the agent's voice terminal and a screen pop of the e-mail, and the user interface, is delivered to the agent's browser. For the duration of the message call, the agent's busy status is noted as by the Definity PBX.

The application includes several built-in message-handling tools, including the ability for the agent to create an e-mail response, place the message on hold to consult with other agents within the call center, or forward/transfer the message for handling by another agent; for example, a subject matter expert via e-mail (see Exhibit 6-10).

Customers can interact with a call center agent using whichever method of access meets their needs, preferences, or equipment capabilities. Internet telephony allows voice calls to be sent across the Internet, effectively utilizing a single telephone line to handle both voice and data.

Text chat is implemented as part of the call control applet that is downloaded to the customer's PC when a call is initiated. It supports the voice session between the customer and the agent, or it replaces the voice session for customers who do not have a multimedia-equipped PC. For example, the applet provides a convenient means of accurately transmitting strings of digits, such as account codes or serial numbers, and of verifying spoken communications — "Did you say 'B' or 'P'?" — as well as a method of access for the hearing impaired. A similar applet is downloaded to the agent's desktop and provides the text chat interface.

Customers choosing e-mail as their preferred means of communication with a business via the Web typically wait long periods of time (several days

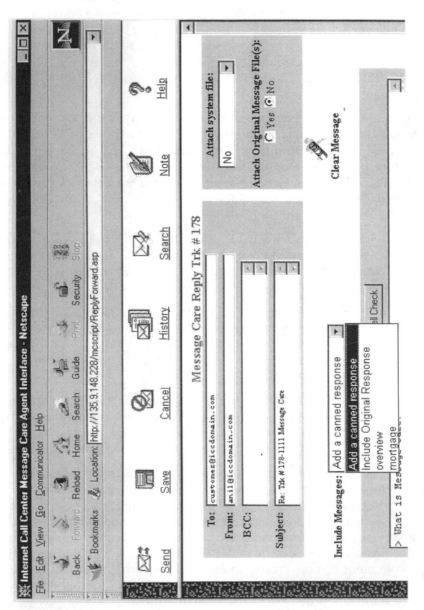

Exhibit 6-10. E-mail management and customer response screen.

to several weeks) before receiving responses, leading to dissatisfaction. Lucent's package provides these customers with direct access to call centers, ensuring a guaranteed response. A customer simply clicks on an e-mail link on a business Web site, then submits his or her request using a form provided by the site. The request arrives at the desktop of the first available, appropriately skilled agent via a screen pop. The agent composes the response, preferably with the help of a database of frequently asked questions, and then sends an e-mail message back to the customer.

Free-formatted e-mail is supported in addition to form-generated e-mail, allowing the call center to support general inquiries from customers who contact the business using a standard Internet address — not just the hot links to a Web site. For greater flexibility, customer faxes can also be routed to an electronic mailbox via an e-mail server's fax interface, where the fax messages can be accessed and routed just like Internet calls or e-mail.

In addition, for those customers who have a second telephone line, the software will accommodate callback requests. The customer simply indicates this preference and his or her telephone number. The request is queued and routed just like any other call coming into the call center. The first available, appropriately skilled agent will then automatically return the customer's call using the PSTN. The agent is provided with the customer's telephone number within a pop-up window, allowing the agent to verify the number. The agent then selects "Call" to signal the PBX to automatically dial the number and connect the call.

For the customer who has a firewall that will not permit Internet telephony access to the desktop, a page is "pushed" to prompt the customer to enter callback information. The customer submission is routed to the first available, appropriately skilled agent for callback.

CALL CENTER PHYSICAL DESIGN

A mundane, yet important part of work force management is the design of the physical environment. Over time, call center managers have found the following to be useful enhancements to the workplace:

- Keep walls between employees below 50 inches.
- Use indirect lighting. Full-spectrum lighting (that simulates natural sunlight) is best.
- Use ergonomic (adjustable, comfortable) chairs.
- Make sure monitors swivel and can be set at or below eye level.
- Control sound. Use carpets, sound-absorbing foam, acoustic wall paneling, and soft-surface tiles on the ceiling.
- Use plants and pictures to soften the general demeanor of the office.

CALL CENTER TRENDS

Call center technology reflects the CTI industry in general. As computing begins to dominate telephony, the industry begins to take on the turbulent character of the desktop world, where new and old players rise and fall with sinusoidal regularity. Some near-term industry trends include:

- Diminutive call centers. Smaller call centers are more practical. The physical infrastructure needed in the past (all agents in one place) is no longer necessary, enabling agents to be geographically distributed. Another factor driving down the economic size of call centers is the proliferation of open systems and "un-PBX" solutions (using an NT box, ActiveX controls for software, etc.).
- Virtual call centers. With call center technology available at the network level, agents can work from home or regional offices.
- Blending of inbound and outbound agents. More robust software allows agents to participate in both activities.
- Integration of measurement. Internal statistics and external (customer survey) statistics.
- Empowerment of agents. With better software tools and desktop information, agents can monitor their own work more closely.
- Outsourcing of overflow traffic. Rather than staff to the peak loads, overflow can be routed to third-party call centers or permanent part-time agents.

SUMMARY

Call centers are the engine of change in the telephony industry. As they move beyond voice and integrate communications from many other media, their position as a revenue generator will become even more prominent. Many of the concepts developed through years of experience in the large call centers can be used on a smaller scale. Even an organization that has no need for a call center may have one or more Help Desks. The principles of good call center design apply to a five-agent Help Desk as well as a 600-agent airline reservation center.

Chapter 7
Preparing the Request for Proposal (RFP)

THERE ARE MANY REASONS WHY ORGANIZATIONS ELECT TO INSTALL NEW TELEPHONE SYSTEMS. Visible failures (such as losing dial tone during business hours) or frustrations experienced by senior management may serve as the engine of change. In other cases, the business may be serving a new market that requires call center capabilities. Regardless of the initial impetus for a new system, the months spent in selecting vendors, services, and equipment are an excellent opportunity to define needs and develop a coherent telephony architecture.

The sections below outline key steps to develop a robust RFP (request for proposal). In addition, criteria and suggestions for selecting consultants to assist in RFP development are included. Many of the concepts apply to smaller scope decisions as well as major PBX, voice mail, and CTI acquisitions.

REQUEST FOR PROPOSAL VERSUS REQUEST FOR QUOTATION

If the organization is clearly committed to a particular vendor and equipment architecture, an RFP may be unnecessary and even counterproductive. For example, if the vendor of choice is Nortel, and the quantities of phones, digital ports, analog ports, adjunct processors, etc., are known, then an RFQ (request for quotation) is all that is required. An RFQ simply lists quantities and model number, with a request for a price quote from the vendor. Full-scope RFPs may cost a vendor $200,000 to $500,000 for a large installation (if it is a close contest, requiring heavy sales presence). If the organization knows, with certainty, that vendor X is the only solution acceptable, then an RFQ — to that vendor only — may be appropriate. In some cases (e.g., governmental organizations), competitive bidding is required. The key point is that avoiding a costly RFP can reduce the time spent by both parties and *should* reduce the bottom-line price to the customer (assuming that avoidance of RFP costs is used as a negotiating chip). "Going through the exercise" has little value and can alienate unsuccessful bidders who (rightly) feel they were simply used to drive down the price, with no hope of obtaining the business. Given the constant rise and fall of vendors in the telephony business, today's unsuccessful bidder may be tomorrow's industry leader — it pays to maintain good relations with all vendors.

On the other hand, if the organization has an open mind and is considering a significant architectural change, an RFP is the vehicle that allows the bidders to use their expertise to develop alternative designs, prices, and service offerings. The well-designed RFP does not say, "I need 5000 digital telephone sets, each with two B channels and one D channel." Instead, it says, "I need 5000 digital telephone sets that can simultaneously transport voice and data..." In other words, a good RFP states the business requirement and gives bidders the opportunity to suggest solutions from their repertoire of technology. A well-crafted RFP is a balancing act — too many detailed, technical specifications and it becomes an RFQ; too few specifics and it becomes vague and the comparison of the bids becomes difficult.

RFP PREPARATION

Educating the Selection Committee

Even if a consultant writes the RFP, management needs to obtain background knowledge to ensure that the scope and direction of the document matches the strategic direction of the business. Following is a list of techniques that can help the selection committee develop a sense of what is possible and available in the market.

- Site visits. Vendors are only too happy to bring potential customers to visit existing customers with installed equipment. The key is to make sure that the trip is structured and includes written questions, discussions with end users, telecommunications technicians, telecommunications management personnel, and demonstrations of hardware/software. Exhibit 7-1 is an illustrative list of questions (actual questions will vary with each site visit). Answers and impressions should be recorded in a "memo to the file;" after the fifth trip, some of the facts begin to blur unless they are in writing. Ideal site visits are to organizations of similar size, business objectives, and technical environment. When looking for call center equipment to serve 50 agents, a visit to an Amex call center of 1000 agents is not particularly useful.
- Phone calls. Telecommunications personnel are often willing to share information, particularly if questions are well-defined (e.g., "Do you allow your employees to use TTI to move themselves and if so, does that result in any database administration problems?"). Calls to other user organizations can provide a wealth of unbiased information.

Exhibit 7-1. Illustrative list of questions for telephony systems site visits.

1. What are the size parameters for your organization? Number of employees, extensions, digital versus analog, number of floors, buildings, etc.
2. What volume of minutes do you incur per year from your IXC vendors?
3. What is the configuration of your network?
4. What PBX equipment do you use? When was it installed, and how has the vendor performed in terms of service?
5. What voice mail system is used, and how many ports are available for employee use? Do you broadcast messages?
6. What are the general features of your voice mail system?
7. What CTI or IVR applications are installed? What vendors do you use on these systems for hardware and software?
8. What has been your downtime experience over the last few years?
9. Is your e-mail and faxing integrated with your voice mail system? Please describe how the interfaces work (particularly with respect to directory maintenance).
10. Do you network voice mail across several buildings or between cities?
11. Has your current telephony vendor been proactive, and have they brought new technologies to your attention?
12. What critical functions does telephony support? Do you have any "cannot fail" applications and, if so, how are they configured (e.g., redundant PBXs, disaster recovery options from the LEC)?
13. Do you have international operations? If so, has the vendor been able to supply maintenance and equipment at a reasonable price to your international locations.
14. Do you have call center operations? If so, what software/hardware platforms support those operations? What reporting is provided via the vendor's equipment?
15. Do you outsource any functions to the vendor? If so, how responsive have they been to end users?
16. Does the vendor's "culture" fit your organization's culture?
17. Have you received comments from your customers regarding the telephone services? Have they been favorable?
18. How does the vendor respond to any requests for training of your in-house staff? Do they seem to want to "keep it all in-house," or are they willing to share their knowledge base?
19. How easy is your system to administer in terms of moves, adds, and changes?
20. Are you satisfied with reporting of problems (hardware/software) out of the switch?
21. Can you get directly switch data easily for *ad hoc* reporting, or is export difficult/time consuming?
22. Does the switch or voice mail have to be taken out of service to install upgrades or do maintenance? Are you a 7×24 operation?
23. Is the vendor responsive with spare parts or is delivery delayed because of parts unavailability?
24. Does the vendor offer seminars/executive briefings?
25. How well does the PBX and voice mail integrate with the call accounting package?
26. Is the telephony system ready to integrate with the Internet and the IP (rather than circuit-switched) network?
27. How effective is remote monitoring? Have you been notified proactively of impending problems and thus been able to resolve before they became service affecting?
28. Do you feel the vendor has a strong "vision" into the future of telecommunications? Do their products reflect current trends (e.g., open versus proprietary, IP enabled, speech recognition ready, Web-based internal reporting)?
29. Is the vendor able to meet the needs of your small, medium, and large offices equally well?
30. Would you recommend your vendor(s) and equipment configuration to others?

- Web searches. Using a search engine, specific (and sometimes unbiased) information can be retrieved. For example, entering the search term "Open IVR from Nortel" resulted in 164 matches using www.hotbot.com.
- Books and magazines. Telephony books and magazines, although not nearly as numerous as those on data communications and information technology, are growing rapidly. Online bookstores such as www.amazon.com and www.barnesandnoble.com allow keyword searches.
- Research services. Gartner Group, Meta Group, and others provide a complete research service. Many will do a custom analysis of vendor products.

Organizational Participation

Key employees within the business units of the organization must participate in developing the requirements of the RFP. Only business management can address questions such as:

- What drives the decision — cost, specific features, absolute zero downtime?
- What funding is available?
- What functions or features are optional and what are essential? For example, some voice mail systems have software links that integrate directly with the Lotus Notes desktop; others integrate with Microsoft Exchange. How important is this? Is the organization willing to allow yet another interface on the desktop?

Consulting Expertise

If management has in-house telephony expertise, then consultants may not be required. However, frequently either the expertise or the available manpower to do the analysis is missing. In that case, consultants may be required. Following are some of the qualifications that should be examined before engaging such expertise.

- Industry expertise. Experience in switching equipment, CTI, IVR, etc. is critical.
- Ability to simplify and explain. Telephony is steeped in acronyms that obfuscate the business functions provided. A consultant must be able to understand and explain to management. It is not unusual for companies to spend millions of dollars on equipment and not fully understand what they have bought. A translator and explainer is needed.
- Financial analysis. A fair comparison of offers from bidders requires multiple skills. Ports, cabinets, growth factors, hours of voice mail storage, fault redundancy, etc., must be arranged in a spreadsheet so that lifetime costs of each bidder's offer can be evaluated.
- Lack of bias. Consultants sometimes get most of their expertise having worked with one vendor's equipment and software. They know it well

and may have an unconscious bias toward that vendor. The résumé of any consultant should reflect experience with multiple vendors.

Format of the RFP

Appendix A contains a sample RFP. The contents of an actual RFP will vary widely according to need, but will likely include many of the following topics:

- statement of intent
- description of business environment
- a vision statement (the communications environment envisioned by the organization)
- schedule of events
- format of the response, including submission, media, documentation requirements, and caveats
- evaluation criteria
- vendor qualifications
- specific technical requirements
- respondent optional section (bidder has a "blank sheet" to write any comments or suggestions not addressed elsewhere)
- financial options

The "Rehab" Option

One option that should be examined is the choice of simply adding onto or modifying the current environment. For example, the Lucent Conversant IVR server can be connected via analog ports to an older PBX. If an organization just needs an up-to-date IVR and the basic dial tone functions of the PBX are adequate, buying a new IVR server may be sufficient — and would be vastly less expensive. Phone mail systems work the same way. The newest voice mail system can usually be connected to an older switch (although some integration is lost). Some older PBXs, particularly the CO class switch such as the Nortel SL-100, are "tanks" — they very rarely fail. To further decrease the risk of failure, a well-stocked spare parts kit ("crash kit") can be maintained on premises so that critical components can be replaced quickly.

EVALUATION OF RESPONSES

Evaluation of the bidders' responses requires expertise and much effort on the part of the organization's evaluation committee. Because of wide differences in architecture, service capabilities, "just around the corner" feature enhancements, and packaging of software, an apples-to-apples comparison is difficult. Nonetheless, going through the exercise usually makes key features clearer and highlights the need for further information.

There are both quantitative and qualitative factors to be evaluated. Both should be documented to ensure that the evaluation process is fair and as objective as possible.

Quantitative Factors

Some of the objective and measurable factors that can be reviewed and compared include:

- Market share. A low market share is a danger signal. Below a certain percentage, the vendor may not be able to stay in the business long term. This applies even to large vendors, who may offer voice communications equipment through a subsidiary. Of course, a startup may offer an attractive price. This is a risk that management must evaluate.
- Percentage of revenues spent on research and development. Firms with low R&D may not be competitive in the future. Again, it is not the total vendor R&D but that percentage applied to communications. Siemens, for example, manufactures PBXs but also manufactures large power generation equipment.
- Financial stability. The organization's financial analysts should review the vendor's profitability, outstanding debt ratios, etc.
- Equipment mean time to failure.
- Feature comparisons. A matrix of features should be developed for vendor comparison. Having the most features does not always imply the best product, but it does indicate a vendor that is sensitive to market needs. Feature comparison can, however, be misleading when too much emphasis is placed on obscure features that are rarely used.
- Number of "feet on the street." Some vendors have their own service personnel. Others have agreements with various service providers in order to provide domestic or international coverage. Having more direct service employees is sometimes seen as an advantage because of potentially greater quality control over the work.
- Financial analysis of the offers. See below for a more in-depth discussion.

Qualitative Factors

John B. Cotter, in his book *The General Managers*, notes that many of the most important managerial decisions are not amenable to quantification. In fact, many MBA programs devote a large percentage of the curriculum to the relatively small percentage of business cases that lend themselves largely to numerical analysis. Telephony platform selection conforms to this pattern — while "running the numbers" is essential, the core decisions are often influenced by qualitative factors. The following list reflects judgmental issues that might affect selection of the final candidate.

- Product knowledge by the sales representative. Offerings by the major vendors are vast. Documentation is delivered in libraries of CD-ROMs.

Sales representatives must understand the basic functions of their firm's portfolio and have a good relationship with their own product engineers so they can respond quickly to questions.

- Energy level. It is this author's experience that telecommunications vendors vary greatly in their energy, responsiveness, and attention to detail. Some vendors over-rely on price and ignore quality or technology issues. Simple items, such as returning phone calls, preparing comparative analysis of models and offerings, developing designs for disaster recovery scenarios, working with long-distance and local carriers to ensure a smooth upgrade/transition to a new platform, and other activities depend on a sales force that can "hustle."
- Long-term relationships. Most large organizations have departments with different telephony needs. The vendor should be willing to take the time to meet with many individuals in order to fully assess the needs. One group may need the latest technology with "bullet proof" redundant processors, while another simply needs low cost. Months and perhaps years are required to understand the informal and formal decision makers and their needs.
- Ability to deal with a decentralized environment. Most vendors would prefer to deal with one individual or department that can make decisions for the entire company. That is often not the case today. If the vendor is frustrated by having to deal with many decision makers, it will be even more frustrated (and probably ineffective) in dealing with operational issues after the sale.
- Competitive offerings at low, medium, and high ends of the market. It is ideal to have a uniform platform across the organization. For example, NEC has an integrated line — the NEAC1000 IVS/VSP for the very small office (e.g., 10 employees), a midsize NEAC 2000 IVS, and a large-capacity PBX NEAC 2400 IMX (for up to 60,000 ports in one logical "Fusion©" system). It is particularly difficult to establish uniformity of platform when one segment of the line — for example, the 25-port PBX/communications server — is out of line with the rest of the market. The selection committee needs to have a good reason why the little office in Porter, TX, needs to spend 200 percent more on a PBX than on comparable Centrex service.
- Willingness to educate. It is in the vendor's best interest to continually educate key decision makers. Amazingly, some overlook this golden opportunity to present new technologies and concepts to their customer base. For example, few vendors have gotten the message out about voice recognition — the algorithms and DSPs (digital signal processors) have dramatically improved in recent years. Voice recognition enables many previously manual functions to be automated. The vendor should offer seminars, user groups, demos in an executive briefing center, and one-on-one discussions to keep all parties up to date.

- Sufficient headcount. Telephony vendors need to have sufficient employees in the organization's geographical location to ensure that both sales and support are adequate. Depth of technical talent should be present — how many level 2 and 3 support personnel are available?
- Personality. It is a fact of life. In sales, personality matters. If the selection committee does not have good "chemistry" with the sales representative(s), the vendor should put someone else on the job. The RFP process is no place for psychological counseling or remedial sales training.

Financial Analysis

A well-designed financial analysis attempts to answer the following question. All other factors being equal (the economist's *ceteris paribus*), which vendor's solution has the minimum present value of cash outflows over a given number of years (usually five to ten years)? The difficulty in preparing the supporting work papers is to decide what function or feature in one vendor's offer is equivalent to another vendor's offer. For example, one vendor might offer ports that are dual function, that is, they can simultaneously support digital voice and analog modem transmissions; whereas another vendor may have single function ports. If the organization uses few modems (perhaps because of increasing use of the intra/Internet), should the ports be given equal weight? One vendor may offer an extraordinary price on telephone instruments for the initial purchase but revert to a less discounted price in the future. Another vendor may not offer as deep a discount up front but will guarantee discount levels for the next five years. Assuming growth over five years is unknown, the present value of guaranteed discount levels is impossible to calculate. Yet another example: vendor A responds to an outage after business hours in one hour whereas vendor B responds in 90 minutes; how is the difference calculated in monetary terms? Finally, the financial effect of consolidating voice and data and IP telephony over the next five years is not clear.

One frequently used measuring stick is cost per station (total costs, including trunks, equipment depreciation, maintenance, monitoring, and local but not long-distance services will range from a low of $20 per instrument per month to a high of $60 per instrument per month). This is easily understood even by those with little telecommunications background. Also, it can be compared to the cost of Centrex services. Unfortunately, it does not take into account advanced features found in IVR and CTI, nor does it account for quality/redundancy found in fully redundant systems. Organizations that want dial tone, voice mail, local and long-distance access — and *only* that — can obtain these base services very cheaply. The resale market is an excellent source for discounted, older PBXs that can supply base services. Use of refurbished telephone instruments is another popular means of reducing costs. When looking at Centrex, it is important to recognize that each service has a charge (e.g., moves, adds, changes); Centrex charges can be significantly higher in a volatile environment.

Following are some of the factors that should be considered in the financial analysis.

- Amortization period. Possible values range from three to ten years. Some organizations amortize smaller value equipment over three years and larger investments (e.g., more than $1 million) in the five- to ten-year range.
- Moves, adds, changes. Newer communications architectures greatly simplify MAC activity. Instead of changing cross connects, employee telephones are moved via software-only changes. This will significantly reduce MAC staffing requirements.
- Cost avoidance. When making a purchasing decision, it is important to consider the costs that may be incurred if the purchase is not made. Will some business units within an organization purchase their own small PBX to obtain vitally needed functionality? Will more console attendants need to be hired because of growth of business, and can a new system offset some of that growth via technology such as voice recognition?
- Rent. Will the new system have a smaller footprint and thus free up physical floor space?
- Discounts. Do discounts apply to both domestic and international locations? Are they guaranteed into the future and, if so, for how long?
- Equipment maintenance. Is the first year of maintenance covered in the purchase price? What coverage is provided (7 days a week × 24 hours a day), and what is the response time?
- Maintenance personnel. Is an on-site technician provided? If so, what is the level of that individual? Does the vendor supply a substitute when the primary technician is on vacation or sick?
- Software maintenance. What is the definition of maintenance (very important): does maintenance mean fixes or patches to software releases but not new releases or enhancements? Many organizations are surprised when they receive a bill for a new release of software on the PBX, IVR, or CTI server. They assumed that being on maintenance meant all software upgrades would be covered. The evaluation committee needs to nail this down in writing.
- Safety valve clauses. Organizations do not always know precise numbers of ports, users, voice mail boxes, etc. in their current system. There should be a "safety valve" clause that allows excess inventory to be returned and additional inventory to be ordered at the same discount level so that good-faith, quantity estimate errors do not penalize the customer. For example, if 9500 telephone instruments were

125

Exhibit 7-2. Financial analysis of bids from three hypothetical vendors.

	Vendor A	Vendor B	Vendor C
Annual Ongoing Costs:			
Labor (in-house and contractor)	$ 1,900,000	$ 1,900,000	$ 1,900,000
Maintenance, equipment	$ 200,000	$ 190,000	$ 250,000
Maintenance, software	$ 10,000	$ 90,000	$ 110,000
Charges for moves, adds, changes	$ 158,000	$ 250,000	$ 100,000
Anticipated growth (new equipment and software)	$ 84,000	$ 98,000	$ 97,000
Rent	$ 50,000	$ 55,000	$ 49,000
Local exchange and IXC trunking	$ 850,000	$ 850,000	$ 850,000
Subtotal of ongoing costs	$ 3,252,000	$ 3,433,000	$ 3,356,000
Five year present value of annual on-going costs at 9 percent discount rate	$ 12,649,146	$ 13,353,173	$ 13,053,670
Initial investment (one time):	$ 4,500,000	$ 4,600,000	$ 5,100,000
Annual Cost avoidance:			
Business losses due to downtime on old PBX/Voice mail	$ (500,000)	$ (500,000)	$ (500,000)
Additional console operators, salaries	$ (90,000)	$ (110,000)	$ (100,000)
Additional employees required to respond to customers	$ (250,000)	$ (300,000)	$ (300,000)
Subtotal of annual cost avoidance	$ (840,000)	$ (910,000)	$ (900,000)
Five-year present value of annual cost avoidance at 9 percent discount rate	$ (3,267,307)	$ (3,539,583)	$ (3,500,686)
Net present value of alternatives (PV of ongoing costs + initial investment + PV of cost avoidance)	$ 13,881,839	$ 14,413,590	$ 14,652,984

Conclusion: Based on this analysis, Vendor A's proposal is the least cost over five years.

ordered but only 8900 were needed, credit should be given for the returned 600 instruments. There is an equal risk with software. It may be determined, after looking further at the LAN/WAN/e-mail environment, that a particular package may not be a fit. There should be a clause to allow exchange or credit for the value of the software ordered.

Exhibit 7-2 shows a hypothetical financial analysis of offers from three vendors. Cost avoidance is included because it can affect the decision. For example, if one vendor provides a superior IVR, it might help the customer reduce the number of employees that must be hired to meet new business demands — and more so than the other vendors.

SUMMARY

The RFP is the visible part of the iceberg. If organizations wish to avoid the fate of the White Star Line's *Titanic*, all the assumptions under the surface need to be understood. For the respondents to do their best job, the specifications must be based on the most realistic requirements (both qualitative and quantitative) possible. Key members of the organization will need to spend time and money to become sufficiently informed. Fortunately, if done properly, the RFP becomes the strategic blueprint of the organization's communication direction and crystallizes support for the project.

Chapter 8
Implementing Telephony Systems

THE SUCCESS OF IMPLEMENTING TELEPHONY SYSTEMS — whether large-scale PBXs, IVRs, or CTI applications — depends on detailed planning at the organizational and architectural levels. This chapter presents implementation steps for a PBX installation, but many of the concepts apply to telephony projects in general.

End users are paying attention on implementation day; any failures, regardless of the source (hardware, software, local telco, IXC vendor, or failure to educate the user), will make a lasting negative impression. Future projects and funding will be imperiled and users may not give the system a second chance if the promised features do not work. In short, it is essential to plan, test, retest, educate, perform risk analysis, and provide back-out plans if the worst occurs.

The chapter sections below use the model of a large-scale PBX, voice mail, and CTI applications implementation for several thousand users. However, the same principles and concerns apply to smaller projects. A seamless implementation requires an almost obsessive attention to detail. A telephony project manager cannot have too many checklists.

Although the scenarios presented below do not represent a single, actual implementation, they have been pieced together from a number of projects that the author has reviewed or participated in. With technology and business evolving so rapidly, implementation surprises abound. The words "I did not know they [the users] were doing that ..." are heard frequently during implementation week.

THE PROJECT TEAM

Exhibit 8-1 shows a typical project team organization. The usual caveats apply: roles should be clearly defined, at least some team members should have detailed subject matter knowledge, users should be included in an advisory role, and second-, third-, and fourth-level expertise from the hardware/software provider(s) should be available for consultation.

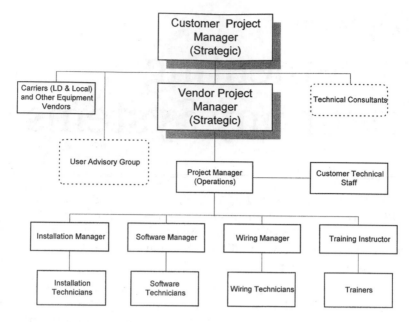

Exhibit 8-1. Project team organization.

THE USER ADVISORY GROUP/IMPLEMENTATION COMMITTEE

The User Advisory Group is a critical part of the project team. It serves as the vital link to the people using the services (and funding the project). Some duties of the user representatives include:

- Approval of implementation dates. For example, a commodity trading organization may have one particularly heavy trading week in a month. The marketing department may have a special promotion on a weekend that will result in thousands of telephone calls coming in through an 800 number.
- Financial review. Not all users will want all features — communication of costs is essential so that capital can be used most effectively.
- Review of technical architecture decisions. CTI, IVR, and voice mail decisions can affect e-mail, directory services, LAN utilization, etc. As telephony becomes more closely integrated with information technology (particularly IP networking) across the organization, decisions cannot be made in isolation. For example, the telecom group might prefer to see Lucent's Message Manager used by all users to consolidate their voice mail, e-mail, and fax. The IT group, on the other hand, may have developed their architecture around Lotus Notes as the consolidating desktop messaging architecture (i.e., all audio clips/voice mail, fax, and e-mail would be presented to the user via the standard

Notes interface). Lucent's Message Manager provides some features (such as playback and recording) that are not included in Lotus Notes. The disadvantage of using Message Manager is that users may not want yet another application on their desktop. Decisions with such broad impact need to be reviewed across the organization.

- Review of features. The telecom group should simplify the feature set (take acronyms out and explain the functions in business terms) and review with the committee.
- Approval of system parameters. The committee should review security parameters and other global attributes.
- Monitoring vendor performance. Along with the telecom group, the committee needs to evaluate progress of the project and ensure that vendor performance is adequate.
- Encourage participation in user training.

SURVEY OF THE CURRENT ENVIRONMENT

Unless the system to be installed is going into a new office with new users, a detailed inventory is essential. Particularly if a telephony system has been in place for many years, there will typically be undocumented hunt groups, special processes, etc., that will not be immediately visible. Without a strong investigative effort, these hidden applications and processes will be found at the worst time — immediately after the cut to the new system, when all available resources are devoted to immediate problem resolution.

Following are components of an enterprise telephony system that should be inventoried. Some of this information may have been obtained earlier in an RFP process, but the numbers must be as accurate as possible for implementation.

- number of digital stations
- number of analog and special-purpose stations (e.g., operator console, call center agent, multimedia telephones, etc.)
- ACDs
- hunt groups (watch for hunt groups no longer needed or with pointers to extensions that are no longer in use)
- circuits, local and long distance; include all types (ISDN, T1s, analog lines, DID, CO, emergency backup, etc.) Ensure all carriers are included. Detailed circuit information should be obtained. For example, indicate whether circuits are loop start, ground start, etc. for IXC (interexchange carrier) trunks.
- 800 numbers
- protocols for ISDN (e.g., National 2 or custom specifications)
- special circuits (data connections going through the switch, videoconferencing nx64, etc.)

- tie lines to other locations
- T1 connections to other equipment, including another PBX or an ATM device. For example, some trading organizations may use a central PBX for the majority of back office employees but use a trading switch (such as the IPC Turrett system) for the traders. The trading switch may get dial tone from the general user PBX and thus will have T1 connections.
- campus fiber connections or other in-house circuits
- connections to peripheral devices such as a Cisco Stratacom ATM box
- wireless connections to the switch
- recording equipment (e.g., Nice Log recorders)
- directory applications
- paging functions (outcalling or overhead paging)
- CTI applications (screen pops, dialers, call center applications such as looking up checking account balance, etc.)
- IVR applications for business functions plus any ancillary functions such as turning on building lights after hours
- peripherals (hardware) and number of ports for each
- a listing of all scripts (so they can later be re-recorded if necessary)
- facilities information: what power is available? What weight load will the flooring (raised or otherwise) bear? Will there be an interim period when old and new equipment will be present?
- floor-by-floor, office-by-office blueprints to enable installers to more efficiently place telephones
- voice mail statistics to determine hours of storage, number of ports, call volumes, etc.
- call accounting peripherals
- equipment room specifications (power, space, wiring, etc.)
- existing Help Desk, ACD, reporting tools
- all toll-free numbers and their termination points (DNIS)

In addition to the above, an inventory should include special emphasis on wiring. The old saw that, "It is in the wiring, stupid" is not far off the mark. Wiring problems occur because there are so many locations where it can be disrupted — cables coming from the digital or analog ports, the MDF (main distribution facility), the IDF (intermediate distribution facility), horizontal cables, and station jacks. Factors to be included in the implementation plan include:

- Main distribution frame. If a new MDF is required, the work should start early in the project because it will house the leads coming from the switch. The MDF should be clearly labeled so that technicians can easily identify cross-connect locations. A decision will need to be made regarding punch-down block sizing. Two sizes are common: the more prevalent 66 type block and Northern Telecom's Bix block.

- Riser cable. In a multistory building, cables are run from the MDF to the wiring closets (IDF) via riser cables. These thick cables must pass through flooring and be appropriately labeled. In buildings with existing wiring, it is possible that there will be insufficient space in the holes between floors to allow riser cables to pass. In that case, additional borings are required (and engineering studies to ensure there are no structural problems caused by holes with a wider diameter).
- Cat 3/Cat 5 wiring. Given the small increase in price of category 5 cabling versus category 3, the higher bandwidth Cat 5 should be used in new construction. Cat 5 will carry both voice and data, at speeds above 100Mbps. There should be a quality inspection of the connections to ensure that the ends are not unwound to make it easier to install. Unwinding causes degradation of bandwidth and increases error rates. Voice-only requires Cat 3 but with the continuing integration of voice and data traffic, it is shortsighted to install anything less than Cat 5.
- Potential to use extra pairs. Some cabling plants have unused pairs. In situations where an older PBX is being replaced, the unused pair can be made "hot" and the older pair used for the existing PBX. This allows for a parallel implementation in which both PBXs are functional for a period of time. New telephone instruments can be placed along side the older ones to give users an opportunity to become familiar with the new system. This technique only applies when new instruments are part of the implementation (note that digital sets almost always use a proprietary protocol and thus are not interchangeable between vendors; the protocols are usually a minor variant of ISDN, but are just different enough to prevent interoperability). A flash cut is the most common implementation — existing pairs are moved to the new PBX and no parallel is feasible.
- Cable management. Standards for installation should be established, including cable management. Good installs include wiring harnesses, complete documentation, appropriate lengths of cables, and high-quality patch panels. The decision needs to be made early in the project as to whether a formal cable management system will be developed. Otherwise, data may be entered manually (or in a spreadsheet) and later imported into the cable management database.
- Terminate, test, and tag. This is stating the obvious. However, it is such a major effort that it should be clearly stated in any project work plan and contract — each riser cable and station pair should be tested prior to implementation. Any new cables must be appropriately labeled.
- Physical access. Technicians and specified others should have keys, codes, and other access permissions to perform testing/installation as needed.

- Environmental problems. Any locations with asbestos or other hazardous materials should be treated with appropriate measures.

Non-Stop Applications

Some telephony applications cannot be shut down or may be shut down only for minutes. A good plan provides workaround solutions such as:

- 800 number redirection. A CTI application can be run at another location by having any incoming 800 number calls redirected to an alternate location during the conversion.
- Centrex can be used for what would normally be a DID number. The local telco may have a disaster routing option that directs a number or series of numbers from one location to another (e.g., all calls that come in for number 281-358-1234 are now directed to 281-321-3456).
- IXC reroute. For example, AT&T could reroute, at the network level, calls going to a specified number.

STATION REVIEWS

A successful implementation requires detailed interviews with users. The starting point is call coverage — how does a call get processed? For example, employee #1 may receive a call and then that call transfers to voice mail if she does not pick it up. Employee #2 is a vice president whose calls are routinely answered by her administrative assistant. If the administrative assistant does not answer, two other assistants may "pick" the call. Finally, if there is no one in the call path, the call is transferred to employee #2's voice mailbox.

The above scenario is straightforward. When call centers, Help Desks, or other more complex telephony environments must be transitioned to a new system, considerable time is needed to understand the user's requirements and make system changes to accommodate those requirements. For example, assume a large Help Desk department allows agents to make personal calls but requires the agent to put any personal calls on hold and answer a business call immediately. In many vendor's ACD software, an agent cannot be "available" and also be making another call (whether personal or not). A crude solution would be to provide agents with 2500 (analog) type telephones side by side with their business station — but that takes up a lot of desk real estate and requires extra sets. A better solution might be to provide a subtle alerting light (or button) that indicates a business call is coming in so that the agent can put a personal call on hold and pick up the business call. The key point is that there are many different scenarios that need to be reviewed and fitted into the new system. Without the requisite advance work, there will not be enough time to make corrections during or after the conversion. If available, a printout by extension from the current PBX, voice mail, directory, etc., would also be useful as a guide.

End-User Training

Without a strong end-user training program, the switch implementation will be considered a failure, even if it is technically flawless. Particularly in lean organizations, users may skip training, assuming that they can get by with a small "cheat sheet."

There are two keys to success: (1) an internal advertising campaign to increase awareness, and (2) a plethora of opportunities for training through a variety of media, such as CD-ROM or video.

Following is an illustrative list of techniques to enhance user understanding of the telephony system.

- Create initial awareness by e-mail or voice mail from a high-level executive within the organization. The message to employees should include dates of implementation, a schedule of training dates, and a list of benefits to be obtained from the new system.
- Assign someone to coordinate and plan the training classes. This becomes a significant effort when several thousand users need basic instruction.
- Contact the building/facilities group early so that appropriate classroom space and cabling can be completed before classes are scheduled to begin.
- Avoid starting classes too soon before switch implementation (users will forget if they do not have the opportunity to practice relatively soon after training).
- Use as many media as possible. For example, training can be provided on video, CD-ROM, intranet sites, paper-based documentation, and IVR ("for instructions on how to save voice mail messages, press 1, etc."). Stand-alone kiosks have also been used to show continuous video demonstrations in areas of high employee traffic (such as the company cafeteria).
- Motivate attendance with low-value but unique gifts. Toy telephones, T-shirts, and other items that are not available except through the training classes can encourage the employee who has a marginal interest in attending a class. Gifts should be drawn into the service of training in every way — T-shirts, for example, could have basic telephone set and voice mail instructions printed on them.
- Develop multi-tier classes. Training levels should fit the needs of the employee. For example, general users may only need a 45-minute class; attendant console operators, Help Desk personnel, and call center agents may need three hours of instruction. Forcing a tax accountant to sit through several class hours, discussing such items as hunt groups and class of service restrictions, is probably inappropriate.
- Set up lunchtime demos of the equipment.
- Develop a matrix showing commands on the old system versus the equivalent on the new system.

- Set up a list of executives and key employees that should be given instructions one-on-one.
- Set up a second and third line of defense. Some employees may be absent during the weeks preceding an installation. Instructions should be placed on every desk before users arrive after cutover. Telecom employees with brightly identified shirts can walk through each department so that users can ask questions on the spot. The Help Desk should be manned with additional staff for several weeks after an installation.
- Consider making training mandatory. A senior management directive may be the most effective technique for encouraging attendance.

BUILD THE DIAL PLAN, CLASS OF SERVICE, AND ROUTING TABLES

These tables require careful thought and coordination with the user community because they affect the way users dial. They also influence the risks the organization incurs from toll fraud and other security threats.

The dial plan (a table or series of tables in the PBX) tells the switch how to interpret dialed digits. For example, when the user dials 9, the system immediately begins routing the call to an outside (not internal) trunk group. Also, the dial plan determines the length of certain dial strings (e.g., all dial strings that start with 2 are four digits long, and any dial string starting with 6 must be three digits long and correspond to a trunk access code). The dial plan contains all the extension ranges that can be used within the system and all those that can be dialed.

An ideal dial plan can be used across the entire organization. Users can use the same dialing scheme to dial the headquarters office, regardless of what other office they are dialing from (as long as they are within the organization's network).

The class of service/class of restriction tables set up the capabilities of each extension. For example, international dialing, long distance in general, or even local service can be disabled for a particular class of service (e.g., for a lobby phone). Some organizations define a class of service zero that does not allow the telephone to dial any number at all. One might question the value of a "do-nothing" COS. However, it can be useful if the telecommunications group is trying to identify modem extensions that may be unused. The extension is set to COS = 0. If there is a user complaint, the COS for that extension can be quickly set to a standard value. It is much faster to do that than to rebuild the extension in software. If there are no complaints for several months, then the analog port can be safely reassigned because no one has been able to use the extension while it was set to COS zero.

A sample COS table is shown in Exhibit 8-2. By intelligently designing the class of service values, there will usually be a sufficient number of classes to match job requirements for all an organization's users.

Exhibit 8-2. Sample class of service table

	0	1	2	3	4	5	6	7	8	9	10	11	12	13	14	15	16	17	18
Auto Callback	n	n	y	n	n	n	y	y	n	n	n	y	n	n	y	y	y	n	y
Call Fwd-all calls	n	n	n	y	y	n	n	y	y	y	n	y	y	y	n	y	y	n	n
Data Privacy	n	y	y	y	n	n	n	n	n	y	y	y	y	y	y	y	y	n	n
Priority Calling	n	y	y	n	n	n	n	n	n	n	n	n	n	n	n	n	n	y	y
Console Permissions	n	n	n	n	n	n	n	n	n	n	n	n	n	n	n	n	n	n	y
Off-hook Alert	y	y	y	y	y	y	y	y	y	y	y	y	y	y	y	y	y	y	y
Client Room	n	n	n	n	n	n	n	n	n	n	n	n	n	n	n	n	n	n	n
Restrict Call Fwd-Off Net	y	n	y	y	y	y	y	y	y	y	y	y	y	y	y	y	y	y	y
Call Forwarding Busy/DA	n	n	n	n	n	n	n	n	n	n	n	n	n	n	n	n	n	n	n
Personal Station Access (PSA)	n	y	y	y	y	y	y	y	n	n	n	y	y	y	y	y	y	y	y
Extended Forwarding All	n	y	n	n	n	n	n	n	n	n	n	n	n	n	n	n	n	n	n
Extended Forwarding B/DA	n	y	y	y	y	n	n	n	n	n	n	n	n	n	n	n	n	n	n
Trk-to-trk Transfer Override	n	n	n	n	n	n	n	n	n	n	n	n	n	n	n	n	n	n	n
QSIG Call Offer Origination	y	y	y	y	y	y	y	y	y	y	y	y	y	y	y	y	y	y	y
Auto Exclusion	n	y	n	n	n	n	n	n	n	n	n	n	n	n	n	n	n	n	n

Automatic route selection (ARS) is used for calls that go outside the organization's premises to the public network. Automatic alternate routing (AAR) performs a similar function but connects only to the private network.

EQUIPMENT READINESS AND ROLLOUT

The service provider responsible for installing the hardware should have standard, detailed checklists to ensure that all the parts are present and "burned-in" well before the installation date. Some typical equipment install and testing steps include:

- Do a box-level inventory. Has everything been received?
- Install cabinets.
- Connect the components (cabinets, peripherals, etc.).
- Test interswitch connectivity (if more than one switch is present).
- Test power; power up the switch(es).
- Cable out the switches and related peripherals.
- Run digital cross connects on the frames.
- Install basic inbound call distribution package.

SOFTWARE INSTALLATION FOR THE SWITCH

Software loosely includes core operating systems, application-level software, call processing, scripts, and any custom voice recordings that must be entered or translated from a predecessor system.

Typical steps include:

- Install operating system.
- Install auto attendant and call processing.
- Set up non-DIDs for switch internal activities.
- Build dial plan, class of service, ARS, and other tables.
- Set up all system parameters (often developed on a PC and then downloaded to the switch).
- Build trunk groups.
- Add any special applications that reside in the switch itself (e.g., ACD groups).

ADJUNCT PROCESSING

Adjuncts encompass CTI, IVR, voice mail, and all the other peripherals that give telephony power beyond simple switching functions. In the frenzy of a large installation, adjuncts may not receive the detailed attention they deserve. General installation steps for adjuncts include:

- links between the switch and voice mail; also, intra-server links if voice mail requires more than one server (e.g., Lucent Hicap software)
- Paging interfaces
- Links to other telephony systems (e.g., a primary switch may provide an analog dial tone to another switch via internal T1 trunking)
- Telecommuting links (extender cards and devices)
- Calling accounting/SMDR
- ATM, multiplexer, and other communications boxes that link to the switch
- Music on hold
- Audioconferencing bridge
- Recording equipment (e.g., to record legal conversations or for security)
- Fax servers
- Environmental adjuncts that control building heating, air conditioning, and lights

SET UP HELP DESK

During a large-scale implementation, the Help Desk receives perhaps 10 to 50 times the normal volume of calls. If it is a difficult installation and some problems remain unresolved for several days, users may call repeatedly. To handle the maximum number of calls and keep track of the requested changes, a special-purpose "transition" Help Desk should be set up well in advance of the implementation. Some special factors to consider include:

- Consider staffing 24 hours a day during the installation. Some users may call at night or call from international locations. Inevitably, some users will have missed training.

- Have as many trained people on the Help Desk as possible. The overall impression of the transition will be based, to some extent, on the speed with which individual problems can be resolved.
- Ensure that equipment is ready for a high-volume Help Desk operation: plenty of ports into the switch, all the right permissions granted, PCs and software available, extra telephones, and a special hotline so that telephony management can direct the implementation team to work on critical problems. The manager or director of the group should not "get in line" with other users when requesting a fix to the CEO voice mailbox. A Centrex line may be appropriate for this function.
- Organize station reviews and trouble tickets so that they can be quickly checked off and reviewed. Users may call several times and technicians may be sent twice for the same problem. The more structured the trouble ticket lists are, the less duplication of effort will occur.
- Establish categories of urgency in advance. For example, nonfunctioning equipment takes priority over a display name change (e.g., Bob Smith wants his name to show as Bob rather than Robert).
- Keep hourly statistics on the number of calls received, the current backlog, and the number resolved. Such statistics will help determine the need for additional resources.

PERFORM A PREPAREDNESS REVIEW

Just prior to installation of the system, the team members should participate in a preparedness review. This allows all components of the cutover to be considered as a whole to ensure that testing and available resources are adequate. Some topics to address include:

- Have all cable pairs been toned, tested, and tagged at the MDF (main distribution frame)?
- Have networking and trunking reviews been completed (including capacity planning)?
- Is the inventory of equipment final? Will there be a sufficient quantity of stations (including specialized equipment, such as special stations for ACD agents, wireless telephones, etc.)?
- Have the systems been certified by the vendor(s) responsible?
- Has the long-distance carrier(s) fully tested all new circuits?
- Has the local carrier(s) fully tested all new circuits?
- Has the user community been repeatedly notified of the implementation? Have arrangements been made for non-stop applications?
- What risks have not been considered?
- Have telephony testers been used to simulate volume dialing in order to stress the environment in test mode?

DETAILED CUTOVER PLAN

During the conversion period, human resources are strained — sometimes to the limit of physical endurance. Anything that can be done in advance helps. The cutover plan is a detailed, chronological listing of steps to be taken by each party during the "cut." It can be written to an hour-by-hour level of detail, including "go/no go" points and "point of no return." Events may change what actually occurs, but starting off with a detailed plan ensures that all parties have a concrete understanding of their responsibilities and backout plans if the conversion fails.

Cutover procedures would typically include the following steps.

- Listing of organizational responsibilities. Names, organization, telephone numbers (work, home, cellular, pager), and other information are required. All vendors, telephony personnel, users involved in the project, and senior management (for escalation) must also be included. Develop an hour-by-hour work force commitment.
- Key functions. Telephones, special applications, etc., that are priority should be noted so that special attention can be devoted to maintaining their uptime.
- Escalation procedures, including names, titles, and phone numbers.
- Equipment installation procedures. Technical implementation steps.
- Network installation procedures. How and when to cut circuits as well as internal links (such as fiber between campus PBXs).
- Load and double-check DID ranges for local and IXC circuits.
- Critical meeting times and locations for go/no go decisions.
- Provide adequate floor space and clerical support needed during cutover.
- Establish freeze dates and times for changes.
- Establish date and hour to start.
- Set up emergency procedures.
- Evaluate any switch translations to ensure that tie lines, etc., function properly.
- Test consoles.
- Test SMDR interfaces.
- Test voice mail networking (AMIS, VPIM, and proprietary links).
- Test method for users to get old voice mail for a period of time.
- Install and test any special software (ACD, CTI, etc.).
- Inspect spare parts kit.
- Test 800 number termination.
- Test all tie lines (including access codes).
- Test overhead paging.
- Test executive areas.
- Ensure compatibility of adjuncts with new equipment (e.g., PBX to audio conferencing equipment).

- Make final port assignments and jumper work.
- Provide monitoring and review by expert control center (diagnostic center).
- Provide last-minute training for temporary help, technicians, etc.
- Immediately after cut, back up all critical files on the new system.

BACKOUT PLAN

Obviously, an organization cannot function without telephone service (or, for some organizations, even some of the services beyond dial tone). Thus, a backout plan is essential. A recovery plan should include at least the following elements:

- quick reversion of system parameters and all settings to pre-cutover mode
- relink adjuncts and peripherals to pre-cutover mode
- reconnection of lines at the MDF
- verification that critical user services have been restored
- restoration of inbound and outbound local and long-distance carrier facilities
- retest all circuits
- test stations (analog and digital)
- communicate status to all parties, including plans for future cut

SUMMARY

Implementation of telephony systems requires more detailed planning than many IT projects because of the exceedingly high expectations of users. Using traditional project management techniques, along with assistance from vendors and carriers, can go a long way toward a smooth implementation.

Chapter 9
Voice over IP

MOST FUTURISTS AND OBSERVERS OF INDUSTRY BELIEVE THAT INFORMATION AND NETWORKING herald as significant a change to society as the introduction of fixed-location agriculture or the introduction of coal- and gas-powered machines in the nineteenth century. A subset of this revolution is the transition from analog/waveform-based communications to digital/packetized transmissions. When communication is digital, networking resources can be shared, optimized, and monitored far more efficiently than in the analog world of a "channel per conversation." George B. Dyson, in his book, *Darwin among the Machines* (Perseus Books), elegantly states the impact of this transition:

> A circuit-switched communications network, in which real wires are switched to connect a flow of information between A and B, would be swamped by the intractable combinatorics of millions of computers demanding random access to their collective address space at once. All the switches in the world could never keep up. But with packet-switched data communications, collective computation scales gracefully as the number of processors (both electronic and biological) grows... Consensual protocols, running on all the processors in the net, maintain the appearance of robust connections between all the elements at once. The resulting free market for information and computational resources determines which connection pathways will be strengthened and which languish or die out. By the introduction of packet switching on an epidemic scale, the computational landscape is infiltrated by virtual circuitry, cultivating a haphazard, dendritic architecture reminiscent more of nature's design than of our own.

Voice over IP (Voice over Internet Protocol, sometimes abbreviated VoIP) is in the relatively early stages of deployment worldwide. However, the economics are obvious to all the communications vendors (both carriers and hardware manufacturers). The first decade of the twenty first century will see the rapid transition of all networks (voice and data) to a digital/packet-based architecture. The chapter sections below outline the structure and implementation options currently available for VoIP.

FUNDAMENTALS OF IP TELEPHONY

Communications across a network come in two flavors: circuit switched (connection oriented) and packet switched (connectionless). Circuit-switched networks (still by far the most common for voice) form a dedicated connection between two points. In contrast, packet-switched networks segment data into small pieces of information called packets (also called datagrams). Because the packet does not have a physically pre-destined place to go, it must have an address to tell the network hardware where to forward it. In a connectionless architecture, there is no call setup. Each packet traverses the network independently. This was one of the original criteria of ARPANET, the precursor of today's Internet. The network was to remain functional even in the case of nuclear war; hence, packets must be able to get from point A to point B in multiple ways.

IP networks in general are favored by most organizations because they

- offer a large base of vendor support (enhances interoperability)
- operate within the framework of many operating systems
- are highly scalable
- work over a variety of underlying (layer 2) media such as ATM, Frame Relay, dedicated lines, ISDN, Token Ring, and all varieties of Ethernet.

Exhibit 9-1 shows a simplified IP telephony system built from the traditional PBX and circuit-switched architecture and an IP gateway. Conceptually, IP gateways function as follows:

- A T1 or analog line connects from the traditional, circuit-switched PBX to the gateway.
- A user dials a telephone number as usual. The PBX (via its routing table) sends the call to the IP gateway.
- The IP gateway has a table that converts the telephone number into a specific IP address.
- The gateway then sends the call, in packetized form, over a private IP gateway (dedicated circuits, Frame Relay, etc.) or possibly the public Internet. Typically, the telephone signal is highly compressed to conserve bandwidth (much of human speech — at least *polite* human speech — is characterized by pauses that do not need to be transmitted). Compression of 8:1 or 16:1 is common.
- On the receiving end, the process is reversed. The packetized signal is converted back into a circuit-switched call (i.e., one call per channel) at the receiving gateway and sent to the destination PBX where it is connected to the proper user, based on DID telephone number (or the call could go into an automated attendant, IVR, etc.).
- The gateways are typically full-duplex, so that conversations can be carried on both ways simultaneously. An example IP telephony system using an IP gateway with an existing PBX infrastructure is shown in Exhibit 9-2.

IP telephony was introduced in 1995 with the introduction of VocalTec's Internet telephone software. Release 5 of their Internet Phone, targeted to home and SOHO users, includes call conferencing, video-conferencing,

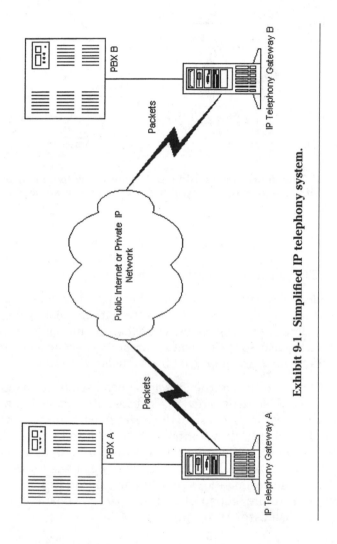

Exhibit 9-1. Simplified IP telephony system.

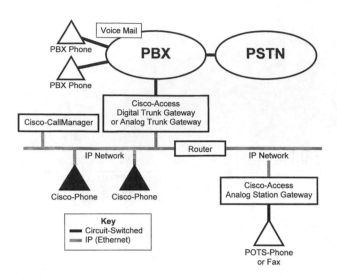

Exhibit 9-2. IP Telephony system using an IP gateway with an existing PBX infrastructure (Courtesy of Selsius Systems, www.selsius.com).

Internet telephone calls to others on the same system, whiteboarding (to share and edit photos, documents, and drawings online, realtime), and calls to the PSTN network via a gateway (fee-based).

Some service providers are setting up gateways in major cities around the world so that they can offer low-cost international dialing via an IP network. The advantage of using gateways is that the massive investment in the PSTN over the last 100 years can be preserved — the same telephone and dialing techniques are employed by the user.

PC-to-PC voice over the Internet has also been used by individuals to by-pass toll charges. Most medium to large organizations have not been interested in this architecture because of the strict reliance on the Internet, the requirement that the user do his or her own special addressing, the proprietary nature of the solutions (must have same equipment at other end), and simply the fact that the other end needs to be online for the call to go through. Some of these limitations (particularly the addressing issue) may be resolved in the future, but strict PC-to-PC voice appears to be more for hobbyists than businesses at this point.

Voice Quality Considerations

One of the limitations of sending voice signals over an IP (data) network has been the quality of the sound. Because data networks were not originally designed to send packets at a constant rate, the sound at the other end can become garbled because all the packets (some of which may have traversed multiple paths in the network) will not arrive at the same time. In a data network, "bursty" traffic does not typically interfere with the operation of the application. In addition, packets may be lost in a data network and reconstructed later by retransmission or reconstruction. This architecture sometimes results in latency (delay in receiving the message). Latency is noticeable when it exceeds about 0.25 seconds. Users of satellite telephones are familiar with the disturbing delay between speech and transmission at the other end.

Industry developments are reducing the delay by:

- more powerful chips and DSP boards in gateways
- compression of voice traffic using a voice coder (vocoder) (The newest devices, called parametric vocoders, have the characteristics of the human vocal track programmed into their chips and transmit only the data needed by the algorithm. The newest vocoders produce near-toll-quality speech.)
- use of private networks where traffic loads can be controlled, rather than the public Internet
- use of protocols that guarantee quality of service (QOS) (these protocols allow voice and other streaming communications media to have guaranteed or reserved bandwidth so they are not interrupted. This will be discussed in more depth later in this chapter.)

BENEFITS OF VoIP

As noted above, there are many advantages of voice over data networks. Some benefits visible to the user include:

- Reduced or eliminated toll charges because the traditional Telco network is bypassed.
- Use of the same telephone line for voice and data in small offices and homes. Realization of this concept may take several years.
- Increased functionality available to users. Using digital, packet-based telephone systems allows the user to more easily integrate the desktop into telephony functions (via graphical user interface). The traditional POTS telephone has only 12 buttons — with no backspace.
- Internet features. Internet Call Manager from the Canadian firm InfoInterActive is a Web-based telephony product that illustrates features that can be enabled via IP/Web technology. As one would expect for the home/SOHO market, it works with Windows 98 and Windows NT. Following are options available to a single-line home or SOHO user.
  - Call accept. The calling party hears a message stating that the called party has seen the call and will accept it shortly. The Internet session is logged off and the software transfers the call to the telephone line; the Internet session can be resumed afterward.
  - Call transfer. Calls can be transferred to a second line or a cell phone.

147

- — Call acknowledgment. The caller can be notified that the called party is on the Internet. A message can be left.
- — Voice mail. If the called party does not respond, the call goes to voice mail and can be later retrieved locally or remotely.
- — Call logging. Calls are logged with time, date, calling name, and calling number. Note that caller ID is not required for this service to work. The software must be supported by the individual's ISP provider (e.g., AOL).
- Click to call services. By adding HTML code to its Web site, an organization can provide its customers with a convenient way to click a button on the Web page and get a phone call. No premise equipment is required for this service. See Exhibit 9-3 for an example.
- When voice is carried by the LAN, significant savings in both hardware and MAC (move, add, change) costs can be obtained. See later section on LAN-based telephone systems.
- Enterprise applications can use LAN servers to act as programmable switching and routing nodes. Phone service can be delivered to employees that move frequently in a building (or warehouse) by combining wireless LAN services with IP switching.
- Future benefits. Although not yet deployed, two new technologies will likely be in the marketplace in the next few years: (1) realtime display of costs during calls, and (2) more integrated voice recognition, along with voice signatures for authentication and enhanced voice mail applications.

ARCHITECTURE OF IP TELEPHONY GATEWAYS

There are a plethora of IP telephony switches and gateways on the market today. Some are designed for small offices to front end an existing (circuit-switched) PBX. Others may serve as a carrier-grade IP telephony system that can use the traditional PSTN as well as IP networks.

When reviewing gateway options, the following functions should be considered.

- Is the gateway H.323 compliant?
- Does it support a high-compression CODEC based on standards such as G723.1, GSM, G711?
- Does it support real-time faxing over the Internet (using the T.38 standard)?
- How many ports are supported? Will it easily scale up?
- Does it require a "line side" T1 or can it use an analog interface?
- Is the gateway housed in a fault-resilient chassis? Does it have a fully redundant and hot swappable power supply, cooling fan, and mirrored hard drive? Does it have appropriate alarms, temperature monitoring, remote notification of problems or remote restart?

Instant live response by telephone

Ring-Now procedure explained

Introducing the first on-screen Ring-Now button for Internet home pages — the fast, easy way to create sales, improve services and turn browsers into customers.

This technology uses the InVADE telephony platform on a CT Server to create phone calls between a Web browser and the companies owning the browsed Web pages.

contact

* Please fill in all fields marked with an asterisk

From which country are you calling? *
`United Kingdom (no: Loncon)`

Please enter the telephone number on which we can contact you *

Instant or Scheduled call * `Now`

Click the telephone key pad to Ring-Now

Exhibit 9-3. Click-to-call services (Courtesy of InVade Corporation, www.invade.com).

If an IP switch instead of an IP gateway is used, the following questions apply:

- Can it bill to end users (customers) by transferring data to an ODBC-compliant database, such as Microsoft SQL Server?
- Can multiple IP switches connect to a central database for sharing of rate tables, tax tables, codes, etc.?
- Does the switch include software such as North American numbering plan verification, access verification, destination of phone call detected via DTMF detection, DNIS verification, and CO emulation? Note that CO emulation allows digital trunk lines to act like traditional CO lines.
- Does it perform echo cancellation?
- Is call processing supported, including tone generation, volume control, and call progress?

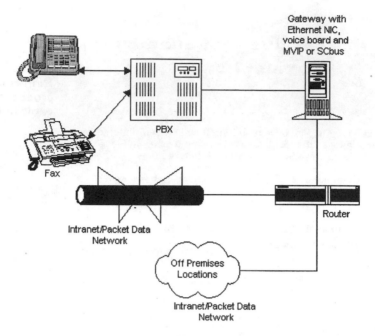

Exhibit 9-4. Intranet voice gateway attached to a local PBX.

VoIP AND FAX USING PRIVATE LINES/INTRANET

Using an existing private network (leased lines, Frame Relay, ATM) is a robust method of transporting voice via packets. The organization can control the quantity of traffic going over the network and can monitor uptime, etc. Exhibit 9-4 illustrates an intranet voice gateway attached to a local PBX.

Fax capability is often included in VoIP gateways. Fax can tolerate delays up to three seconds as a result of spoofing techniques used in intelligent fax boards (e.g., Brooktrout TR114 board). Fax machines in the office may cost $3000 to $7000 per year to operate (total costs, including maintenance, toner, and employee time). In addition, the long-distance charges for fax use are significant — particularly if the long-distance fax service charges for retries where public lines are not conditioned well. Fax modems use the same digital speech processing boards as VoIP so that the customer typically can get a bargain from the hardware manufacturer.

TELEPHONES BASED ON LAN ARCHITECTURE

For many years there have been two communications infrastructures in office buildings: one for local area networks (data) and one for the centralized, circuit-switched PBX/voice/fax/modem network. The emerging market of LAN-based telephone systems uses the data network as its sole network and takes advantage of the myriad of capabilities available in the data world.

The traditional PBX star architecture where all points converge to a single point is well-understood and widely deployed. However, it does have some limitations.

- The entire voice network in an office building is dependent on the functioning of a single (perhaps multiple) switch. If the PBX goes down, all voice communication (except perhaps Centrex or cellular phones) stops. In general, PBX = single point of failure.
- Voice architectures are somewhat clumsy to scale. The organization has to anticipate growth, ensure that cabinets have sufficient shelf space to house new cards, and internetwork PBXs (e.g., Lucent DCS) when a single PBX cannot be further expanded.
- PBX vendors, to some extent, are shackled in the chains of their own history. Because they must support legacy customers, even their newest designs have many proprietary features. Digital telephones, for example, are not interoperable between vendors (except for some generic ISDN handsets).
- Networking is resource intensive. As mentioned earlier, traditional circuit-switched voice communications dedicates a channel per voice conversation. Even with compression, the channel is still tied up as long as the conversation continues. Voice over IP gateways can be used to convert to packet traffic, but this requires the purchase of gateways at each node of the network.
- Tools to monitor the voice network are not comparable to the sophistication of data network tools such as HP OpenView. Any of the major network monitoring tools, for example, can oversee large, globally distributed heterogeneous environments. In contrast, there is no PBX management package that will simultaneously monitor at a detail level the functions/alarms of Lucent, Siemens, Nortel, and NEC PBXs (as would be the case in a decentralized organization with data equipment from multiple vendors).

Using completely IP-based telephone systems (not merely an IP gateway) may solve some of these problems. In this architecture, the users have H.323-based telephones that tie directly into a fast Ethernet LAN (or some other switched/shared media network). The voice/packets are routed via switched hubs. The advantages of this approach include:

- Less proprietary architecture. Telephones that directly tie to a LAN are generic in the sense that they follow the H.323 standard from the ITU-T. (H.323 defines how call control, CODEC, and channel setup occur over data networks that do not offer guaranteed service or quality of service).

151

- Less maintenance intensive. Moves, adds, and changes in a large organization are expensive, in part because wiring from the PBX to a telephone must go through a specific set of ports, cross connects, and wiring closets to get to a fixed location in the building. PCs, on the other hand, can have an IP address that allows e-mail to get to them regardless of their physical location, so long as they are on the network.
- No central failure point. Using multiple Ethernet switches, for example, mitigates the effect of a single failure. It is not an "all-or-nothing" dial tone environment.
- Simpler administration tools. Although some vendors have developed GUI administration tools for traditional PBXs (e.g., NEC AimWorx), managing telephones connected to a data network is more efficient.
- Easy scalability. Because switching is distributed, it is an easy task to add another 100Mb Ethernet switch onto the network. Capacity can be added gracefully rather than via forklift.
- Long-distance charges can be reduced. Same as with the PBX/gateway solution.

The Internet and Frame Relay networks all have the problems of latency and inadequate quality of service. However, LANs have controllable and robust pipes linking nodes in the network. In a LAN implementation, IP telephones are linked to a switched Ethernet hub. Since the switched hub sends packets only to the intended destination (not everyone on the same node, as is the case with older style hubs), intense data traffic is not disrupted by voice traffic. Voice takes considerably less bandwidth than streaming video, for example. A typical LAN-based telephone system is shown in Exhibit 9-5.

Handsets can be H.323 compliant, with a built-in Ethernet transceiver and an H.323 CODEC. Another option (but perhaps not as acceptable to most business users) is use of the PC as a telephone.

In a LAN-based telephone system, there still needs to be a communications server. Some vendors provide a PC-based server in which a distributed set of servers manage switching, call control, routing, and IVR. Another option is to use peer-to-peer IP-based telephones that are H.323 compliant. To get to the outside (i.e., link to the CO), a gateway server is required.

VOICE OVER FRAME RELAY

Voice over Frame Relay is a remarkable feat, akin to Dr. Samuel Johnson's comment about a dog walking on its hind legs — "It is not done well; but you are surprised to find it done at all." Some organizations use Frame Relay successfully; others find the voice quality too rough for business use. However, no one argues that the cost is significantly lower than traditional dedicated circuits (costs as low as $0.01 per minute).

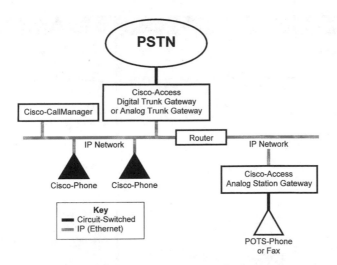

Exhibit 9-5. Typical LAN-based telephone system (Courtesy Selsium Systems, www.selsius.com).

Frame Relay is the offspring of the X.25 Protocol, which was designed for poor-quality microwave and copper transmission facilities. To make a point, one Australian engineer successfully passed data traffic over cotton string soaked in salt water using the X.25 Protocol. X.25 achieves its powerful accuracy over "dirty" lines by checking each and every packet, using a cyclic redundancy check on the contents.

The strength of X.25 is also its weakness. By checking every packet, along with negative acknowledgment to the sender, the delay becomes significant (around 500 ms). This much delay makes X.25 unsuitable for audio, streaming video, or applications that require rapid back-and-forth response.

Frame Relay, designed to operate over modern digital circuits and fiber backbones, is a layer 2 protocol that checks the validity of a frame of data but does not request retransmission if an error is found. Instead, the higher level application is held accountable for correcting any errors. X.25, on the other hand, is a layer 3 protocol and includes flow control, error detection and correction, supervisory information, etc. Because Frame Relay does not do such extensive error checking, the frames are routed far more quickly from one node to another in a Frame Relay network.

Frame Relay supports both PVC (permanent virtual circuits) and SVC (switched virtual circuits). The result is that communication circuits can be shared by many users at the same time (which is why carriers charge less for Frame Relay circuits than for traditional dedicated T1s or fractional T1s). Exhibit 9-6 is a conceptual illustration of a voice over Frame Relay network.

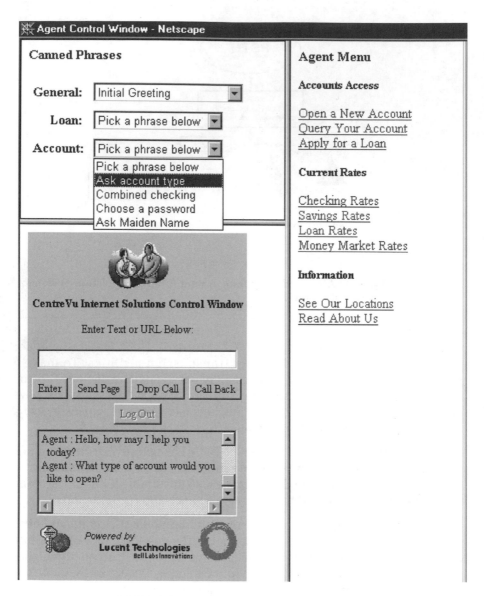

Exhibit 9-6. Voice over Frame Relay network.

Most carriers provide Frame Relay services with a CIR (committed infor-mation rate) that specifies a guaranteed transmission rate of packets. If bandwidth is available, the rate of transmission can go significantly above that rate. Because voice cannot be delayed without severely degrading the quality of the sound, the CIR of the Frame Relay service should be adequate to carry most of the voice traffic (e.g., a 64Kbps CIR). The DE (discard eligible) bit is set to zero for frames that should not be discarded. Theoretically, these are those frames that do not exceed the CIR. Frames above the CIR are marked with the DE bit equal to 1 and can be dropped. Organizations have used the capacity above the CIR in the network as an added benefit because most of the time frames with DE equal to 1 get transmitted. However, as carriers fill available fiber resources, free bandwidth above the CIR is increasingly rare.

Cost Factors

Estimating in advance the cost of a Frame Relay network is time consuming; network managers may become frustrated when trying to compare offerings from vendors. Major cost elements include the following.

- Access line to local telco. Unless a bypass vendor is used, the Frame Relay access line that connects the organization to the public Frame Relay network goes through the local telephone company.
- Port cost. An organization's access line will terminate into a serial port on a Frame Relay switch, located at the frame operator's premises. Various sizes of ports are available — from 64Kbps to T1 speed. Frame Relay traffic cannot exceed the port speed.
- Private virtual circuit charge. This charge to set up the circuit on the provider's network is driven by the specified CIR. The charge is not distance sensitive (sometimes called a postalized rate).
- Frame relay premises equipment. FRAD, CSU/DSU, etc.

Exhibit 9-7 shows some ballpark Frame Relay costs for volume-related parameters. These costs are not based on any specific vendor (because prices vary considerably), but represent order-of-magnitude pricing at the time of this writing. When comparing pricing, determine if the voice over frame algorithm transmits silence/pauses in the conversation. According to the Frame Relay Forum (www.frforum.com), only 22 percent of speech is essential to the transmitted. Of the remainder, 56 percent represents pauses and another 22 percent can be regenerated because it includes repetitive patterns.

Quality of Service and Delay Sensitivity

When designing the Frame Relay network, managers should review frame length-limiting options. Otherwise, long data frames can interfere with voice frames that need to get through the network in a timely manner. This is a subset of the more fundamental problem of quality of service (QoS).

If cost were not an issue, QoS would not be relevant because the CIR could simply be increased to the level where contention would not be important.

155

Exhibit 9-7. Sample pricing for selected frame relay components.

Ballpark Costs for Frame Relay Services

Port Cost	Installation	Monthly Costs
Port size = 56Kbps	$ 325	$ 60
Port size = 1.544Mbps	$ 300	$ 425
Access Cost (Intralatta)		
Port size = 56Kbps	$ 550	$ 45
Port size = 1.544Mbps	$ 550	$ 150
Access Cost (Interstate)		
Port size = 56Kbps	$ 350	$ 45
Port size = 1.544Mbps	$ 350	$ 100
Connections per port		
2–8 connections (each)	$ 13	
9–13 connections (each)	$ 9	

However, cost is a major driver of Voice over Frame Relay (and Voice over IP), and CIR is one of the factors that directly affects total line charges, along with the port charges.

Some providers, such as MCI, offer varying priorities to their subscribers. A sampling scheme is used to prioritize traffic for a specific customer. Traffic between customers is not prioritized. This is not a true QoS and may require specific router hardware at both ends.

Some other techniques/factors for QoS include:

- assurance that abundant bandwidth exists
- reduction in superfluous traffic (for example, multicast broadcasts to every port, regardless of need, wastes resources. Sometimes this is termed "pruning." By defining members of a specific group using DLCIs [data link connection identifier], bandwidth can be conserved. In addition, conference calls can be supported using the same technique.)
- bandwidth planning; make sure that "edge" traffic stays there and only goes on the enterprise backbone if it needs to
- avoiding long frames

Equipment and Configuration Considerations

Real-life implementations of Voice over Frame Relay use equipment with proprietary solutions. Although there is a new standard, the FRF.11 implementation agreement from the Frame Relay Forum, it will take some time for all the vendors to gear up production to meet the new standard for interoperability.

There are a number of compression algorithms used by Frame Relay vendors. Some of the most common include:

- PCM and ADPCM. These are toll-quality compression algorithms but require large bandwidth (64 or 32Kbps).
- ATC (adaptive transform coding). This is a simple system with a variable digitization rate.
- ACELP (algebraic code-excited linear prediction). One of the latest and well-researched compression algorithms, ACELP allows near toll-quality speech over Frame Relay all the way down to 4.8Kbps.

Other factors to be considered when reviewing FRAD equipment include:

- Congestion management. The device should respond to a traffic load by varying queue size before congestion occurs.
- Jitter buffering. Jitter is the variable latency for packets moving through a network. It disrupts the perception of smooth audio; the user hears irritating pops and cracks. Use of a large jitter buffer allows the incoming packets to be smoothed so that users hear continuous speech output.
- Fragmentation. To prevent large packets from disrupting voice communications, frames should be fragmented to smaller sizes (see Exhibit 9-8). For example, ACT Networks (www.acti.com) limits packets as follows:
  — voice: maximum 83 bytes per frame
  — asynchronous data: 71 bytes per frame
  — synchronous data: 72 bytes per frame
  — fax: maximum 58 bytes per frame
- Priority service. By putting data into buffer queues first, fax and voice can be given higher priority to smooth delivery to the user.
- Silence detection. Equipment should be able to detect silence and dynamically give that bandwidth to other channels as needed.
- Ability to set the discard bit on. Because the retransmission of a voice packet does not make sense for a real time environment, any voice packets must be set to discard eligible so that the system does not waste resources trying to retransmit a packet when it is too late to be of value to the user.

INDUSTRY ENABLERS OF VoIP

The communications industry is moving rapidly to develop a more robust architecture to support VoIP. Some of these developments include:

- IP version 6 (Ipv6). The address space of the current version of IP (version 4) may be exhausted within a few years. Although the current 32-bit address space can support four billion devices on more than 16 million networks, the use of classes A, B, and C to define

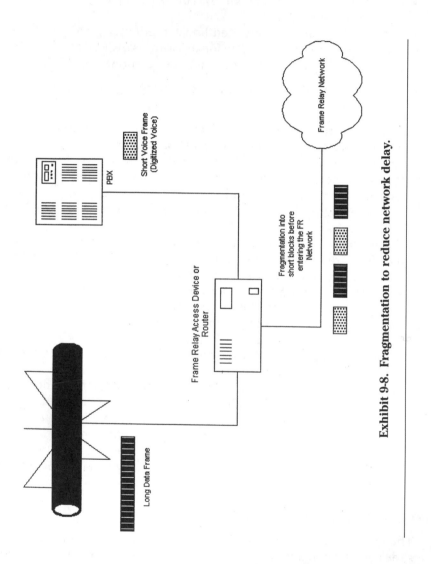

Exhibit 9-8. Fragmentation to reduce network delay.

organizational related networks limits the actual addresses that can be used. So, the IETF has a new standard that will greatly expand the number of addressable devices (128-bit network address). While they are at it, a number of new features are being added. The most important for VoIP is a priority flag (four bits of the datagram) that indicates whether the traffic is congestion controlled or noncongestion controlled. Congestion-controlled traffic can be slowed down to relieve the congestion. Voice (or realtime audio, for example) would be classified as noncongestion controlled because it is delay sensitive. Although this can help in the relative priority of noncongestion-controlled traffic, it does not automatically bump up the priority of all noncongestion traffic above congestion-controlled traffic. Ipv6 will help but not eliminate packet timing problems.

- Winsock 2. Winsock is a network programming interface and functions as the link between Windows operating systems (Windows 98 and NT) and the TCP/IP protocol. What is different from the previous version, Winsock 1.1, is that version 2 includes a mechanism to negotiate quality of service with a network and hence improve performance for various multimedia communications, including audio.

- IP switching. Rather than using routers, which are latency prone, more networks will use IP switching devices, such as Ipsilon's IP Switching technology. These devices route traffic at a lower level in the network stack (at layer 2 rather than the more processing intensive layer 3 for routers). As a result, they are quicker and promote a higher quality of service for real-time traffic. A number of industry observers have noted that, in the long run, the best solution to bandwidth is a dumb but quick network.

- RSVP (Resource Reservation Protocol). This IETF (Internet Engineering Task Force) standard specifies how bandwidth should be reserved across various network topologies. RSVP broadcasts a user's quality of service request to end devices such as routers. Signals are sent back and forth to all intermediate devices along the path, resulting in the desired quality of service all along the line (if the capacity is there to start with). This enables, for example, audio to come before a long file transfer or e-mail attachment. Currently, IP traffic on the Internet moves on a FIFO basis. TAPI 3.0 and Cisco Systems incorporate RSVP in their architectural designs. It may be some time before this standard rolls out across the public Internet, but at least the preliminary steps are being taken by some of the larger vendors.

DEPLOYMENT ISSUES

A few years ago, an editorial was written for a large city newspaper describing all the factors that influence the survival and growth of downtown, urban centers. The author made the comment that average people must have a compelling reason to go downtown, not simply because they feel sorry for it or that they "ought" to use downtown services. Similarly, real network managers must have a bottom-line business case for VoIP that either saves money or improves service *now* — moving to VoIP simply because it is architecturally elegant does not serve the enterprise. Hence, the transition to VoIP should be carefully studied from investment, ROI, and timing perspectives before embarking on implementation. Following are some practical factors that should be reviewed by communications management.

Negatives

- Traditional, circuit-switched PSTN per-minute rates are dropping, especially for dedicated-to-dedicated transmission (office building to office building, where access T1s, OC3, etc., are in place).
- Quality of service issues remain with many current implementations, although there is a frenzy of activity devoted to resolving the problems.
- VoIP requires an investment in capital for gateways and other equipment.
- The VoIP network must be managed and maintained.
- The industry is still in the emerging technology stage; standards need to be solidified before hardware prices are driven down by interoperability and availability of sophisticated equipment.
- VoIP billing strategies have not yet matured. Most organizations do not have the means to properly bill or allocate costs for networks based on packet traffic. Looking at the public Internet, the question of reserving bandwidth runs into the traditional model for billing used by most Internet service providers. ISPs, like the airlines, oversell available capacity and count on customers to accept slowdowns. If, at some point in the future, high-priority bandwidth is to be available for delay-sensitive traffic, how are customers to be billed for the additional costs of higher speed routers, etc.?

Positives

- Standards such as H.323 have been established and vendors are moving to make more of their products compliant.
- International locations still incur such high PSTN charges that sub toll quality voice can be tolerated for the large savings.
- For specific situations, VoIP can have a quick payoff (particularly if VoIP over the Internet can be deployed). For example, a January 1999 article in *Datamation* magazine quoted St. Louis-based Universal Sewing Supply as having a payback of less than one year for a $27,000 investment in VoIP hardware.
- Generally, the same hardware that supports VoIP can, with little cost increase, support fax over IP. Again, international locations benefit the most.

- VoIP usually rides on an existing, already paid for data network. Hence, it is merely filling in the interstices of bandwidth that would otherwise be wasted.
- Traffic can be segmented so that external customers do not see the VoIP network, which is relegated to more quality-tolerant users (employees, contractors, etc.).

VOICE OVER IP TRENDS

The efficiencies of the IP network are compelling. In the long run, circuit-switched networks will be seen as inefficient, costly, and difficult to control. Some of the developments to look forward to include:

- easy to address IP telephones; each person may have their own personal IP telephony address that would be valid worldwide
- replacement of Centrex services with IP functions
- "IP jacks" that allow IP telephones to be plugged in anywhere
- far more flexibility in functions; software could be downloaded from the Internet or telco service provider into an IP telephone; upgrades become downloads
- variable (selectable) quality of service
- much cheaper IP gateways as the price of DSP boards goes down

SUMMARY

Voice over IP is gaining momentum. The economics are compelling because of the enormous bandwidth efficiencies of VoIP compared to traditional circuit-switched networks. Many organizations have existing data networks that could easily tolerate the relatively small increase in traffic from voice. Standards, such as H.323 and RSVP, are enabling organizations to use multiple vendors and drive a much higher quality of sound to the end user.

VoIP may or may not be toll quality, depending on the care and attention that has been given to its implementation. The algorithms are getting better, along with the underlying edge equipment. After determining if the savings merit the investment and attention, an organization should consider a pilot program. Given the lack of maturity of the technology, there should be a thorough test in which equipment parameters, network management, and integration to the existing PBX infrastructure can be reviewed prior to implementation.

Chapter 10
Call Accounting and Telephony Management Systems

TELEPHONY CALL ACCOUNTING IS SOMETIMES REGARDED AS THE "PLUMB-ING" OF THE COMMUNICATIONS WORLD. It requires staff time for development and monitoring, hardware for storage and polling, and a plethora of links to internal accounting systems. Nevertheless, like its real-world counterpart, the business of telephony cannot continue without it, except in the smallest of organizations.

Call accounting is an important component of the entire suite of telecommunications management systems that allows intelligent and efficient management of the organization's voice, data, and wide area networking systems. In addition to call accounting, other management systems include cable/plant management, network management, call resale (if applicable), equipment inventory (warehouse function), reconciliation systems (such as calling card reconciliation), work orders, security notification, and interfaces with Human Resources databases (name, department information).

The following chapter sections outline key concepts and functions for elements of the enterprise telecommunications management system.

CALL ACCOUNTING

Call accounting is typically the first of the telecommunications management systems implemented. It distributes telephony costs to the users and departments that incur the charges. Billings to the user departments include items such as depreciation; moves, adds, and changes (MACs); voice mail fees; and long-distance telephone calls (usually the largest component of the total).

In addition to essential matching of costs to those using the services, call accounting packages provide a wealth of operational information for the telecom manager. In fact, without some rudimentary call accounting functions, a large PBX installation cannot be effectively managed.

Call Detail Recording Systems

The technology to support call accounting starts with the CDR (call detail recording) port off the PBX. Call detail records flow from that port to either a serial RS232 call buffering device (highly recommended) or directly to the call accounting server. Exhibit 10-1 shows a typical configuration of PBX, buffer box, and call detail server.

Call Buffer Box

The quality and capacity of the call buffer device affects the reliability and ease of all CDR. If it runs out of memory, fills up too quickly, does not recognize missing data, etc., the subsequent reports generated will have no value.

Following are points to consider when reviewing call buffer devices (as part of the overall call detail reporting system).

- memory capacity (one megabyte is the minimum [too small for many organizations]; 8 to 32MB is preferable. Roughly 100,000 compressed call records can be stored in one megabyte)
- dedicated serial port for connection to the PBX
- data compression (prior to storage and transmission)
- security function to prevent unauthorized access (both access and commands should have password protection)
- port monitoring for real-time assessment of port activity during data collection
- power protection
- battery backup (because these devices are solid-state, loss of power could mean loss of all data)
- user-programmable ASCII data format for data collection
- speed of dial-up modem
- retransmit if errors occurred during transmission (cyclic redundancy check failure)

Organizations with one or more large PBX sites and many small sites may choose to process call detail at the smaller site by having the call accounting server dial up (poll) a CDR buffer device at each smaller location and download the information. Because PBX ID is part of the CDR record, subsequent reports can be sorted by PBX so that only information relevant to a particular site is presented in each report.

Some vendors have recognized the benefits of using the Internet to transmit CDR data (zero incremental transmission costs) and have developed TCP/IP-based polling/buffers. In addition to reducing the transmission

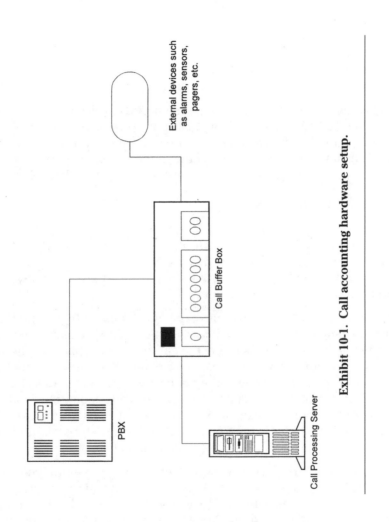

External devices such as alarms, sensors, pagers, etc.

Call Buffer Box

PBX

Call Processing Server

Exhibit 10-1. Call accounting hardware setup.

costs, the IP interface allows the PBX to be monitored by SNMP (Simple Network Management Protocol) management systems, such as HP OpenView.

CDR buffers using TCP/IP (also called Network Pollers) have the following features (in addition to those listed above).

- Ethernet interface for TCP/IP transfer (using telnet, ftp, UDP); 10Base-T and thin Ethernet are common implementations
- SNMP configurable and manageable (i.e., the enterprise network control center — including voice, data, and video — can manage the device remotely)
- alarm recognition (also available on dial-up buffer devices)
- dial-up backup ports (asynchronous I/O ports)

Linking the voice communications system (PBX, voice mail, and adjunct servers) via SNMP (or its successor) to an enterprise-level command center is the right direction. It enables efficient monitoring and brings voice communications into the mainstream of communications. As all voice technologies become less proprietary, SNMP-based monitoring will proliferate.

CALL ACCOUNTING SYSTEM

From the CDR buffer box, the call detail is transmitted to the call accounting server/software. Most of these systems run on a large NT-based server with many gigabytes of storage (a 6000-user PBX, for example, can easily generate a million records a week and three times that many if internal, station-to-station CDR is retained).

The back-end server has a large-capacity database (e.g., Oracle) that houses the millions of call detail records. A number of decisions have to be made when installing and operating the call accounting package (note that some vendors provide an integrated suite of telephony management products, of which call accounting is the major component). In addition, some existing systems may need to interface with the call accounting database in order to provide more meaningful reporting (e.g., names of employees). Using the Telco Research TRU SYSTEM as a model, key steps required to implement and operate a call accounting system include:

- Implement a directory. In order to report on each employee's activity, the system has to associate a name, office number, etc., with a telephone number. The directory contains personal information, organizational levels (department, division, and company number within an enterprise), authorization codes, associated telephone numbers, and PBX ID. In many cases, the organization already has an existing directory that can provide the base for developing the call management directory.
- Cost allocation. The PBX knows nothing of costs — only the duration of the call, the source and terminating numbers, and other technical

information. The call accounting system takes the detailed call detail record and matches it against tariff rates from a carrier such as AT&T or MCI. The cost of the call is then estimated, based on a match between the tariff and the beginning/ending points, using V&H (vertical and horizontal) coordinates. Most organizations have "postalized" rates, meaning that the cost of the call is distance insensitive (for domestic calls) and depends on whether the call originates or terminates from a "dedicated" site with a direct link to the IXC POP, or is a "switched" site that links to the IXC POP via the LEC (local exchange carrier, such as SouthWestern Bell or PacBell). The cost reported to the end user is almost never exactly the same as the cost charged by the long-distance carrier. Special credits, timing differences, and other factors make the reported cost per call always an approximation. One of the options the organization has is to "force" the call accounting system to balance to the long-distance total cost. For example, if the total long-distance bill is $500,000 per month and the total bill per the call accounting system, using standard rates from the long-distance carrier is $600,000, then each call gets reported to the end user as 5/6 ($500,000/$600,000) of the calculated cost in order to force the calculated amount to equal the actual (*pro rata* billing). Some organizations throw in additional charges, such as staffing costs, tie lines, and special projects. Although this is a convenient way to spread telecom costs, it has the disadvantage of showing the user community per-minute costs that are unrealistically high. A manager in a high telephone use department might complain that the organization's rates are higher than she could obtain on the street with another carrier.

Most systems have input screens that allow costs to be defined by time of day. Other screens define user costs (e.g., for a tie line) to be inserted into the database. Other costs, not related to IXC charges can also be included — charges for the telephone handset, routine modifications (such as hunt groups or pic groups), and other one-time charges.

The ultimate reporting potential of any call reporting system depends on the detailed information coming out of the CDR port of the PBX. Exhibit 10-2 lists example CDR fields from a Siemens 9751 PBX.

Note that when the PBX is configured, many decisions need to be made regarding the output of the CDR port. For example, if an admin "pics" a ringing telephone up in his or her "pic group," does the CDR record show the original dialed number or the number of the admin that picked up the call?

Basic Calling Accounting Reporting

Listed below are reporting features that should be included in any call accounting package.

Exhibit 10-2. Call detail reporting example data elements.

Data Element	Example
System ID (identifies the PBX – helpful when several CDR outputs are sorted together in the call detail processing package)	KNOX11
Code for type of call	01 for incoming, 02 for outgoing, 03 tandem (trunk to trunk)
Date the call was made	070799
Time call ended	11:03:09
Duration of the call	Hours, minutes, seconds (note: some older PBXs may have limitations on calls over 24 hours in duration)
Extension that made the call	35027
Number called	713-234-8765
Class of service	05
Trunk number the call came in on	01
Access code dialed	897699
Incoming circuit	987
Outgoing circuit	3456
Node number	1
Attendant console indicator	3
Bearer capacity class (indicates a type of ISDN call)	0 for voice grade, 3 for 64Kbps data, w for wideband
Attendant console	
Condition codes (indicate a variety of events)	0 for an intraswitch call (originate and terminate on same switch), 1 for attendant assisted, 4 for long duration, C for a conference call

- Calls by station (telephone extension). Includes minutes, costs, number of calls, and locations called. Generally, only external calls (incoming and outgoing, local and long distance) are shown for medium to large organizations.
- Departmental summaries, with appropriate subtotals at department, division, organization level breaks.
- Internal calls by extension. Most large (several thousand) offices do not keep internal calls due to the millions of additional CDR (call detail recording) records required. Also, the usefulness of such information is limited because internal calls do not require external trunking. In some situations, however, internal CDR can be useful. For example, if an employee is receiving harassing calls from another employee or illegal activities are suspected between employees, CDR could be used to substantiate the claims (a later section in this chapter will address recording policy and privacy issues).

- Trend analysis.
- Exception/custom reporting. Includes international calling, high cost, long duration, excessive number of calls, and after-hours detail. Also, special numbers, such as 900, 800, operator assistance, etc., are listed for review. Because exceptions will vary somewhat by organization, the parameters should be user definable. Exception reporting helps to identify a pattern of abuse. Generally, any subset of calls should be able to be defined (e.g., calls made after 6 p.m., before 6 a.m., or over 15 minutes to Mexico).
- Geographic summarization. Summaries by area code and city/state/ province can be included. Calls from and to designated locations of interest can be listed in a separate report. Calls to a location that has no business relevance to the organization are probably fraudulent.
- Trunk utilization. This is a key area for cost containment. There should be sufficient trunking to ensure that users are not blocked beyond the specified Erlang blocking factor. Trunking beyond that level represents excess costs. Note: trunking that is adequate to meet the designated blocking factor at peak utilization periods (for most organizations, around 11:00 a.m. and 2:00 p.m.) will be considerably in excess of requirements during off-peak hours. Also, some critical groups within the organization (e.g., medical facilities, traders, key executives) may be configured as non-blocking; that is, there is a 1:1 ratio of extension numbers to external trunks (a channel per telephone). Odd statistics for trunking — zero usage, extremely short telephone calls, or very long telephone calls — may indicate an operational or mechanical problem with specific trunks.
- Toll fraud exception reports. Used in conjunction with long-distance carrier services, this is a powerful tool for protection against telephone hackers.
- Summaries and detail by forced authorization code (FAC). Forced authorization codes, found in most PBXs, require users to enter a code (previously assigned to them) before making long-distance or other toll-incurring calls. FAC is usually based on class of service. Therefore, all users need not be forced to use the code (i.e., the CEO would likely be exempt, whereas any phones in the lobby would require a FAC for long-distance calling). FACs are excellent tools to summarize specific activities. For example, assume a group of outside engineers have been hired to work on a specific project. By giving them all a class of service that requires an FAC, all the long-distance telephone costs associated with the project can be captured for billing to the project.
- Special number alerts. There may be specific numbers that need to be monitored for business reasons (e.g., an employee or contractor suspected of calling the competition and supplying confidential information about the business). These "junkyard dog" reports are useful for all kinds of investigative work.

Some specific pricing features include:

- ability to price based on type of facilities used; for example, the organization may want to designate an arbitrary cost per minute to calls made over tie lines (in order to recover the fixed and ongoing cost of the circuits); private line billing options should also include the ability to charge a fixed amount per call
- breakdown of first and subsequent minutes into tenths of a minute billing
- discrimination of pricing based on time of day (if applicable)
- minimum per-minute charges

Right to Privacy and Telephony Policy

Before implementing call detail recording (in the PBX sense) or call recording (actual conversations of employee's extensions), the organization needs to develop a communications policy. Employees may have expectations completely different from management. Legal positions in this area have been changing recently, but one key element seems to stand out — — employees must be informed about the specifics of the organization's policy. Some questions that should be addressed in policy include:

- Are conversations on company facilities (PBX, phone lines, wide area networks) considered private? In other words, does the company "guarantee" that the conversations will never be investigated or revealed to others?
- Does the organization make a reasonable attempt to keep conversations private?
- Is voice recording performed? If so, are the employees informed (they should be)?
- Are illegal activities allowed? (This sounds like a silly question, but the obvious should be stated in policy).
- Is private use allowed? For example, can employees call their spouses or friends during the day? What is the consequence of calling pornographic 900 numbers?
- How long can messages be retained? Note that in legal proceedings, an organization may be compelled to disclose data or documents related to the litigation in question. This discovery process is broadbased and can include both e-mail and recorded conversations (as well as paper documents). A casual, recorded conversation by an executive — "Can you believe how sloppy the Alpha project was done last year?" — can be most damaging if discovered by the opposing party in a lawsuit. When the court consents to the discovery process, the organization is compelled to make a good-faith effort to provide all material that is relevant. Limiting the duration of stored messages helps ameliorate this exposure.

- How are employees informed of policy? Do they sign an annual acknowledgment statement, indicating they have read and agreed to abide by the policy?

The following excerpt from the Web site of the law firm Schmeltzer, Aptaker & Shepard, P.C.[1] illustrates some of the complexities and hazards associated with voice recording (and presumably even with call detail recording, which only records the event and not the content of the message):

BE CAREFUL WHAT YOU ASK FOR

In a recent interesting decision, the Connecticut Superior Court held that surreptitious taping, regardless of content and location, can constitute a tortious invasion of privacy. The issue arose in a lawsuit against an employee by her employer. The holding in this case has wide-ranging ramifications — regardless of who taperecorded whom.

In WVIT, Inc. *v.* Gray, the court refused to dismiss an invasion of privacy claim brought by television station WVIT and its news manager against a reporter for invasion of privacy. The reporter allegedly taperecorded conversations between herself and her manager (and several other employees). The court held that the fact that the conversations at issue were primarily business related and that they took place in the workplace did not preclude an invasion of privacy claim. Rather, the court determined that the manager and other employees retained a zone of privacy even in the work environment. The court found that while the manager and the reporter's fellow employees had "little reasonable right to expect that business related discussions with fellow employees remain protected from disclosure," they did have a reasonable expectation that these discussions would not be surreptitiously taperecorded by a subordinate.

This case has a number of interesting implications — not only in circumstances where employees themselves engage in surreptitious tape recording to "build a case", but also in invasion of privacy claims — raised by employees concerning employer monitoring of e-mail, voice mail, and other modern technologies. It undoubtedly will also be raised in actions involving desk and locker searches.

Given the climate these days, it is imperative that companies determine in advance what their policy is on access to sensitive areas such as e-mail and voice mail. Employers should then communicate this policy effectively. This aspect of the employment relationship is fraught with peril, and a modest investment in planning and communication can avoid a large amount of exposure in the future.

Cabling and Wiring Management System

Unlike the data world, cabling going to individual handsets is not shared media. Thus, a large office will have thousands of dedicated wires (in contrast to the daisy-chaining of LAN terminals). These wires and connectors are a major cost element of a telephony system and account for many of the service problems. Unfortunately, many organizations have at best an informal system for tracking telephony wiring. Such neglect is understandable, given the considerable start-up effort required to document wires, ports, cross connects, etc. Nevertheless, the benefits in the long run of a cabling and wiring system are clear:

- Technicians do not waste time with trial-and-error techniques to identify the right circuit.
- Identifying available pairs helps reduce redundant wiring.
- Documentation of the cable plant speeds the learning curve for new technicians.
- Planning for expansion and changes is much easier if the location and capacity of existing wiring facilities are known.
- Use of a standard naming technique will reduce errors (particularly for routine move, add, change work).
- Knowledge of current network allows better utilization of unused capacity (extra pairs, etc.).
- Cable vendor performance can be more accurately tracked.

ELEMENTS OF THE CABLE INFRASTRUCTURE

To understand the components of the cable management system, it is helpful to look at the wiring structure of a typical office building. Exhibit 10-3 shows the basic elements of the wiring infrastructure (logical structure).

A building circuit starts at the PBX, travels to the main distribution frame (sometimes just called the frame), where it goes to a punchdown block. The term "punchdown" refers to the manner in which the insulation on the thin copper telephone wire is removed to allow contact with a conductor on the frame — it is punched down into a sharp crevice that cuts into the insulation and makes contact. From the punch-down block on the frame, the circuit travels to riser cable (large group of cabling that goes up and down a building via a conduit). At each floor, part of the riser cable branches off to a wiring closet (called IDF, intermediate distribution frame). From the wiring closet, the circuit travels over individual station wires to telephones, faxes, modems, and other devices that have PBX connectivity.

By having a cabling management system documented with the locations of all wiring, technicians can identify secondary circuits that can be used for new installs or additions to existing offices. If an existing pair can be used, the installation of new cross connects, jacks, etc., can be avoided.

The most common elements included in the cabling database are listed below. Note that each organization must tailor the database to fit the existing infrastructure.

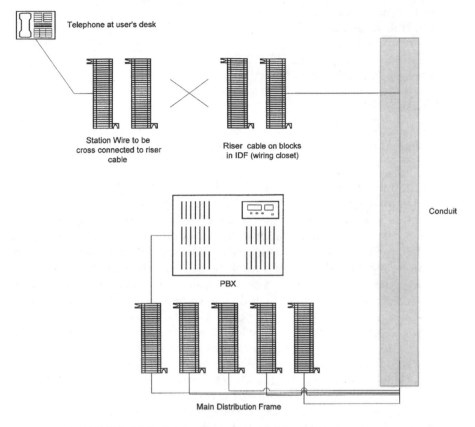

Exhibit 10-3. Basic elements of wiring infrastructure.

- starting and ending locations
- description and function of cables (CAT 5, CAT 3, coax, etc.)
- comprehensive cabling infrastructure, including RS232, Ethernet, Token
- ring, FDDI, voice, ATM, ISDN, and electrical systems
- individual cable lengths (and cumulative lengths are automatically calculated by the system)
- decibel loss (additive loss is also calculated automatically)
- status of the cable run (active, inactive)
- number of pairs
- type of service (data, voice grade, etc.)
- costs (if applicable)
- type of jacks
- common equipment (PBX, key systems, hubs, routers, etc.)
- fiber-optic backbone cables

- fiber-optic station cables
- multiple appearances
- cross connections
- patch cord connections (UTP cables)
- fiber-optic patch cord connections

TYPES OF CABLING

There are obviously many varieties, manufacturers, and specifications for cabling. However, in a typical office building, they fall into the following categories:

- Station wiring. This is the thin telephone wire that attaches to the digital or analog handset on the user's desk. It is often CAT 3 grade, meaning that it will carry voice or low-speed data but not high-speed traffic (e.g., 100Mb Ethernet). Most digital sets now use only a single pair. When pulling wires, it is useful to include a second pair. A second pair can be used for additional functions (e.g., a "stenophone" hoot-and-holler device for traders). A second pair is also helpful when upgrading or changing PBXs — the old system can be run on one pair while the new system can be run in parallel on the second pair.
- Riser cables. Riser cables provide circuit connections within a building. The riser cable is composed of many smaller cables that will spread out in the building. If riser cables were not used, each station would have a separate cable directly linked to the MDF (a homerun cable). Use of riser cables greatly reduces wiring costs.
- Conduit. Usually made of plastic or metal, conduit protects the wiring from environmental hazards and the occasional overzealous technician who might want to take a shortcut. These should be included in the cable management system. One of the limiting factors in any cabling design is the width of the conduit. It provides the means to pull cable from floor to floor, and can also be used to run wiring underground between buildings.
- Cross Connects. For a circuit to be established, obviously every point must be connected. The station wire has to be connected to the riser cable.

GETTING THE CABLING INFORMATION INTO THE DATABASE

Loading the database with correct information (mapping) is a significant effort. A group in the telecom organization (or a vendor) must identify the size of each cable and note its beginning and ending points. Cable pairs must be mapped to a cross connect, frame, or splice. In addition, a complete, formal naming scheme for rooms, IDF and MDF locations, and all relevant equipment must be established.

The following list includes some recommendations for meaningful nomenclature. In practice, it varies according to the organization's time for the project and the complexity of the environment.

- Use standard format for building names, wiring closets, IDF, MDF, rooms.
- Create names that are meaningful; for example, NCC100R (*N*orth building, used by *C*all *C*enter, *100* pair, *R*iser cable).
- Allow for growth.

When setting up the cable management system, a decision must be made: will the database contain end-to-end information at a high level of detail or only at a summary level? For example, "extension 7897 linked to MDF cable ERTV876, pair #34" is less detailed than "Port 98745 connected to ... MDF cable 456 ... WEQW cross connect, row 5, down 10 ... Room 4418 East ... Jack ID A345 ... extension 7897, John Doe." In practice, most organizations elect to use a middle level of detail.

Like some of the new Web sites that provide the browser with driving directions between two specified geographic locations, a good cable management system will suggest the appropriate cross connects between two specified points.

Asset Inventory and Management

Large organizations often lack detailed telecom inventory and management data. As a result, they may lose accountability for equipment, allow warranties to lapse or pay the wrong amount, and be unable to bill internal departments with accuracy. Although it is a large commitment in time and effort, an asset inventory provides a strong tool for management and analysis of the telecommunications infrastructure.

The previous section described circuit and wiring database systems. Asset management goes beyond wiring and includes handsets, PBXs, monitoring equipment, and (potentially) software components such as feature sets (who can do what). Some of the information provided by an asset management system includes:

- equipment on hand and deployed (this is typically user defined and can be everything from stations to PBX circuit packs to punchdown blocks in equipment rooms)
- in-stock availability by warehouse
- warranties — by vendor, due date, equipment serial number
- vendor histories — what percentage of equipment from a specific vendor has been found to be defective
- location (address) of equipment
- original cost

- billing or internal ID that relates the equipment to the owner (sometimes this is simply the telephone extension and table lookups are used to obtain billing responsibility codes [via ODBC-compliant applications])
- soft information such as the feature set and class of service attached to a specific extension (for example, users may be billed for additional features or quantities, such as the capacity to store more than 25 voice mail messages)
- summaries by department or division (e.g., how much equipment was purchased over the last 12 months)
- "shrink" or missing items
- excess or inadequate inventory

Ideally, all the telemanagement systems are seamlessly linked so that costs, monitoring, equipment locations, and data entry are facilitated. Some packages, such as Lucent's TMS (Telecommunications Management Software), are linked to the specific PBX and administration package. TMS is directly linked to the Lucent Definity PBX and Terranova administration software.

Asset management packages for telecommunications vary considerably in scope and reporting capabilities. Most run on NT or Windows and link to other telemanagement modules. The following are capabilities found in major packages.

- relational database
- interface with a problem-reporting package (typically used by Help Desk software, such as Remedy)
- user-defined queries and reports
- user-defined fields for tracking (equipment attributes or objects)
- import/export capabilities
- user-definable billing periods
- ability to handle multiple locations (warehouses)
- link to work orders
- free form "sticky notes" that can be attached to records
- internal security system

The challenge faced by the telecommunications manager is to define and track only those assets that are important. Too much detail can make the system impractical.

SUMMARY

Telephony management, when properly implemented, provides for effective utilization of resources, accurate billing, detailed history by station, organization, or enterprise, and provides proactive trending of resource utilization. There are packages in the marketplace that fit virtually every environment — small key systems to complex call centers. Although implementation requires considerable effort, the long-term effects are reduced expenses and better utilization of capital to deliver services.

Notes

1. http://www.idsonline.com/saslaw/art_tape.htm, July 6, 1998.

Chapter 11
Telephony Security

"Prudent security requires shutting the barn doors before worrying about the rat holes."

Don Parker, *Fighting Computer Crime*

THREATS TO VOICE COMMUNICATIONS FALL INTO THREE CATEGORIES: (1) theft of long-distance services (toll fraud); (2) business loss due to disclosure of confidential information; and (3) malicious pranks. Vulnerability to these threats varies by size and business type. For example, businesses that frequently engage in intense international bidding may find themselves in competition with a government-owned organization. Because the telephone company is often owned by the government as well (PTT[1]), there is a temptation to "share" information by tapping the lines (all it takes is a butt set and knowing which trunks to tap into). While such occurrences are undoubtedly infrequent, they are a threat.

Toll fraud, on the other hand, is ubiquitous. Hackers use stolen calling cards to find a vulnerable PBX anywhere in the world and sell the number on the street (mostly for international calls). Poorly controlled voice mail options and DISA (direct inward system access) are excellent "hacker attractor" features. Medium-sized installations are preferred because they offer enough complexity and trunking to allow hackers to get into the system and run up the minutes before detection. Smaller key system sites do not have the capacity and larger sites often (but not always) have toll fraud detection systems, such as Telco Research or ISI Infortext's TSB TrunkWatch Service.

Two characteristics of the telephone system enhance the hacker's world of opportunity: (1) it is difficult to trace calls because they can be routed across many points in the system, and (2) hacking equipment is relatively cheap, consisting of a PC or even dumb terminal hooked to a modem. Hackers (a.k.a. phone phreaks) sometimes have specific PBX training — they could be a disgruntled PBX technician (working for an end-user organization or the vendor). In addition to their technical background, hackers share explicit information over the Internet (see www.phonelosers.org). These individuals have a large, large universe of opportunity; they hack for a while on a voice system, find its vulnerabilities, then wait for a few months until a major holiday, and go in for the kill. Losses of $100,000 over four days are common. If holes in one PBX have been plugged, they go on to another. In some cases, they use a breach in one PBX to transfer to another, even less secure PBX.

The final category of security break — malicious pranks — gets inordinate attention from senior management, far beyond the economic damage usually incurred. For example, a voice mail greeting could be reprogrammed (just by guessing the password) to say, "Hello, this is Mr. John Doe, CEO of XYZ Company. I just want you to know that I would never personally use any of XYZ's products." Of course, not all changes are minor. A clever hacker that obtains control of the maintenance port can shut down all outgoing calls or change a routing table — there is no end to the damage if the maintenance port is compromised.

The chapter sections below list practical steps that every organization should take to reduce the impact of security breaches.

TOLL FRAUD

Prevention of toll fraud requires unceasing vigilance. For example, NASA and the Drug Enforcement Agency have both been hacked for millions of dollars.[2] The basic steps for toll fraud prevention are detailed below.

Protect Your Maintenance Port. Use passwords of sufficient length (at least ten characters) and change them monthly. This is the absolute minimum protect. Far better is to use a voice recognition system, such as verification systems from T-NETIX, VeriVoice, or Veritel to ensure that only preregistered, authorized technicians have access to PBX and voice mail systems. Switch vendors may balk at the registration process, saying that they must have flexibility to use any available resources to monitor and maintain a system. Do not give in to this pressure. Disgruntled vendor personnel make excellent hackers, and one needs the ability to shut down anyone who is unauthorized. In addition, a system that verifies by voice will also keep a detailed log of entries so that any problems can be traced after the fact (whether related to fraud or merely operations).

Use Common-Sense Calling Restrictions. If the organization never makes calls to South America, restrict the calling patterns to eliminate that possibility. The telephone operators can be given a class of service that overrides that restriction on the off-chance that a legitimate call needs to be made to a restricted location. Calls can be restricted by time of day, day of the week, and location. For example, lobby area telephones should not generally have the ability to make long-distance telephone calls (or at least not international calls). If the organization does not do business on Sunday, restrict outgoing calls on that day. All common area telephones, such as those in lobbies, break areas, and conference rooms, may need to have after-hours restrictions. The mechanism for restricting functions on the

Exhibit 11-1. Illustrative class-of-service parameters.

Feature	Risk Factor
Profile for an executive extension	
Internal calls	None
Local calls	None
Domestic long distance	High
International long distance	High
Automatic camp-on busy	None
Always in privacy	None
Call forward external	High
Call forward internal	None
Camp-on busy	None
Conference call	Medium
Control of station feature	High
Direct call pick up ("pick")	None
Direct trunk select	Medium
Executive override	Medium
No howler off-hook	None
Private call	None
Save and repeat	None
Station speed	None
System speed call	None
Trunk-to-trunk	High
Profile for a lobby extension	
Internal calls	None
Local calls	None
Call forward external	None
No howler off-hook	None
Station speed	None
System speed override	Low
System speed call	None

PBX is the class of service. Exhibit 11-1 lists a typical scheme for class of service in a large office. Many organizations, much to their later regret, have allowed the technical staff to set class-of-service policy. Because the technical staff is oriented toward pleasing the user, there is often escalation over time in the number of users who have the most powerful class of service. In the absence of policy, if a vice president asks a switch technician to give her dial tone capabilities from an international location, the switch technician will most likely comply with the request.

Use Toll Fraud Insurance. Some PBX vendors and most common carriers will provide toll fraud insurance, so long as basic control mechanisms (specified by the vendor/carrier) are in place. Typically, there will be a deductible ($5000 to $20,000) per loss; but at least coverage for the "whopper" hits are covered. The carriers have sophisticated monitoring programs that identify an organization's typical usage patterns and flag unexplained and rapid increases in volumes to particular destinations. Also, some international locations are far more likely to be called by hackers than others (actually, hackers typically sell the "service" to individuals on the street, who then tend to call certain locations more than others).

It is prudent to keep an up-to-date contact list and periodically send it to the vendor (carrier or PBX manufacturer) that is monitoring traffic. For example, assume that an organization is attacked on a Saturday night. The monitoring service identifies hundreds of calls going to Bolivia and Columbia (countries that the organization normally does not do business with) and attempts to call a responsible party on the organization's contact list. If they cannot reach someone in authority, they are hesitant to shut down all outgoing international business because the organization may have essential functions that require outgoing international calls.

Put Tight Controls over Tandem Trunk Calling (going into the PBX, then going to an outside line). DISA — allowing someone to call in, get dial tone, then call out — should be prohibited unless there is some security system in place to control it (such as voice verification). Some organizations will allow calls into voice mail, and then a transfer to dial tone (using a password). Given the ease of password-cracking techniques now available, this service to employees can be expensive indeed. Better to provide them with calling cards for business-related calls outside the office (or an 800-number to dial into, but not out of the office). Sometimes, vendors set up a new PBX and voice mail system and leave backdoor passwords as well as voice mail-to-dial tone capabilities (with only a two-digit password). In smaller locations, the organization will be completely dependent on vendor expertise. When a hacking incident occurs, the maintenance vendor may accept the responsibility or may say that the customer never instructed them to eliminate DISA, etc., *Caveat Emptor.*

Periodically Review Forwarding of Extensions to Dial tone. Any station forwarded to dial tone is "hacker bait."

Educate Operators and Employees to Social Engineering Techniques. One technique widely practiced is for a hacker to call someone and say, for example, "I am from PAC Bell and we are testing your system for some reported problems. Would you please forward me to 9011 so we can complete our trace of the system?" Of course, this transfer gives them dial tone. Another scam is for someone dressed in a delivery company uniform to arrive at the receiving desk with a package for "Mr. X." Mr. X is not there and the hacker asks to use the telephone to call his boss. Apparently, he is put on hold and then gets in an involved conversation with his boss about wrong directions, etc. What he is actually doing is dialing a local number that charges a high per-minute charge for services (e.g., $15 per minute) and gets a kickback from the service provider.

Immediately Request Local Exchange Carrier to Disallow Any Third-Party Charges to the Main Number. Some prisoners, for example, will make long-distance calls and charge them to any organization that allows third-party charges.

Do Not Forget to Periodically Review Call Accounting Reports. Are there calls to a location that the organization has no business reason to call? Some hackers will keep the volume of calls sufficiently small to stay below the radar screen of the long-distance carrier's monitoring algorithms. Sort down minutes called by location and also list single calls in descending order of cost. A quick review can spot problem areas — including some unrelated to toll fraud, such as "stuck" modems.

Educate Users on the Vulnerability of Calling Card Theft. In some airports, "shoulder surfers" observe calling card numbers being keyed in and sell the numbers on the street as fast as possible. Having an 800 number to call back to the office reduces the frequency of calling card calls (as well as reducing the cost). Using a voice verification system to allow secure DISA (see discussion below) also decreases the need for card use. Users, in the interests of expediency, may occasionally give their card number to co-workers. Most carriers, when they detect multiple usage of the same calling card in widely separate geographic areas (e.g., Japan and the United States) within a short period of time, assume fraud. Ensure that all employees who need a card have their own.

Some organizations, concerned about potential misuse by their own employees, contractors, or temporary workers, use prepaid calling cards. The advantage of this technique is that a stolen card number will be used to its limit and then no further charges will accrue. The disadvantage is that it allows for no internal accounting of what the card was used for and sometimes the card is not fully used.

Monitor the Organization's Fax-on-Demand Server. In order to efficiently serve their customers, many firms will set up a fax-on-demand server that accepts a call from the public network and faxes requested information back to the caller. Hackers have recently begun to exploit this service in the following ways.

- Repeatedly calling the fax-on-demand service, asking for faxes to be sent to a 900 or 976 number owned by the hacker (these area codes have a special surcharge associated with them). Of course, the information on the fax is not used, but the minutes accumulate and the calling party (i.e., the hacked party) is responsible for paying the toll.
- Repeat calls to the fax-on-demand service merely to harass the organization by running up its long-distance bill.
- Traffic volume increase by a long-distance reseller who wants to receive greater discounts from the primary carrier (because the cost per minute the reseller gets depends on volume). While rare, it has been known to occur (particularly some start-ups in developing countries).
- Harassment of individuals by sending the fax to a business or residence that did not request it (waking people in the middle of the night, etc.).

One company was hit with over 2000 requests to send a long document to Israel, resulting in a $60,000 telephone bill.[3]

Techniques to detect and defend against fax-on-demand abuse include:

- check the fax system log (or call detail) for repetitive faxes to the same number
- exclude all area codes where there is no reasonable expectation that the organization would be doing business
- exclude area codes associated with high fraud incidence (e.g., 767 — Trinidad and Tobago; 868 — Dominica)[4]
- monitor overall volume of faxes out
- power off and on to clear the queue if it is obvious that the server has or is being attacked
- monitor the fax server over the weekend (particularly long holiday weekends) because that is the favorite time for hackers to start their penetration

Make Use of the Organization's Internal Billing System. If long-distance bills are broken down by department, it is easier to spot unusual activity. Making the internal reports easy to read, with appropriate summary information (e.g., by international location called), provides the organization with more eyes to watch unusual activity.

Use Appropriate Hardware/Software Monitoring and Toll-Restricting Tools. Some features of these tools include:

- selectively allow or restrict specific telephone numbers and area codes
- allow 0+ credit card access but restrict 0+ operator access

Exhibit 11-2. Toll fraud reporting.

Key Information to be Included in Reports

- Authorization codes used
- After-hours activity
- Area codes dialed
- Calls lasting the longest
- Most expensive calls
- Volumes by day of month and week
- Duration distribution
- International calls sorted by country name/code
- Most frequently dialed numbers

Trunk Utilization by

- Individual trunk
- Trunk group summary
- Trunk type summary
- Trunk exceptions

- limit duration of telephone calls in certain areas
- restrict international toll access
- provide for bypass codes
- daily report (sent via e-mail) of suspicious activity, based on pre-defined exception conditions (Exhibit 11-2 shows key information that can be used to monitor toll fraud and employee usage.)

BUSINESS LOSS DUE TO DISCLOSURE OF CONFIDENTIAL INFORMATION

Some organizations have found their bids for projects coming in at just above the competition on a consistent basis. This could be due to coincidence or unauthorized disclosure. It is always a concern when sensitive information is passed over wires or airspace.

Following are some techniques for securing confidential voice transmissions.

Scrambling Device

Use a scrambling device such as the Lucent Surity 3600 (see Exhibit 11-3) or the Motorola KG-95 Trunk Encryption Device (to encrypt at the trunk level for large-scale security systems deployment). These devices, which enable point-to-point and multi-party encryption, protect the conversation from origin to destination (i.e., no intermediate points of clear conversation). Fax

185

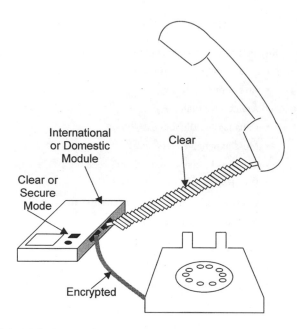

International
or Domestic
Module

Clear

Clear or
Secure
Mode

Encrypted

Exhibit 11-3. Lucent Surity 3600 voice/fax encryption module.

can be protected as well. They typically have a secure/non-secure button that allows the telephone to be used in either mode, as required.

IP Encryption

If the voice conversation is converted to IP traffic before transmission beyond the premises, IP encryption can be used. The Borderguard NetSentry devices, for example, use DES (Data Encryption Standard), 3DES (triple DES), and IDEA (International Data Encryption Algorithm) to scramble any data going across the wire. Note that with the increasing power of microchips, it is much easier for determined hackers (or governments) to break codes. The following quote, found on an Internet security page (http://www.jumbo.com/pages/utilities/dos/crypt/sfs110.zip.docs.htp), illustrates how quickly algorithms once thought secure have become as antiquated as iron safes:

> Use of insecure algorithms designed by amateurs. This covers the algorithms used in the majority of commercial database, spreadsheet, and word processing programs such as Lotus 123, Lotus Symphony, Microsoft Excel, Microsoft Word, Paradox, Quattro Pro, WordPerfect, and many others. These systems are so simple to break that the author of at least one package which does so added several delay loops to his code simply to make it look as if there was actually some work involved.

Enterprisewide Dialing Plan

Use an enterprisewide dialing plan to ensure that all calls go through the least cost and least public route. Calls that go over leased lines (tie lines) are easier to secure than calls going over the public switched telephone network. Encryption equipment can be placed at both ends, and the voice traffic can be converted to IP. Typically, dialing plans are implemented to facilitate ease of use for employees as well as least-cost routing. However, they also increase (at least to some extent) security. A dialing plan is implemented by making changes to every PBX in the organization's network so the user dials the same number to reach an individual, regardless of the location from which the call is made. For example, if Mary Doe's number is 789-1234 and she is located in a Memphis, TN office, then she can be reached from London or Sydney by dialing 789-1234 (with no preceding country codes, etc.); the PBX has all the logic built in to convert the numbers to the appropriate route. A dialing plan also has the side benefit of increasing contact between the telecom staffs of various locations, resulting in an exchange of security information.

Digital Cellular

Convince employees and executives to use digital cellular. Busy executives often use cell phones to discuss confidential topics. There are encryption devices that attach to cell phones (and require a corresponding device on the listening end, whether a land line connection or another cell phone). However, the easiest way to provide basic security is to ask executives to use digital cellular (TDMA[5] or, preferably, CDMA[6]) telephones. The older analog cellular transmissions can be picked up with a pre-1973 television set.

Keep in mind that the U.S. Commerce Department as well as most international governments have significant regulations on the level of encryption used. The French government, in particular, has stringent laws against encrypting without permission.

MALICIOUS PRANKS

Many of the same controls listed for toll fraud will help reduce the exposure to destructive changes by hackers. Some basic prevention steps include:

- Force changes of voice mail passwords. Most current voice mail manufacturers maintain a history of changes so that users cannot change their password to one number and then quickly change it back to the same number they have used for the last ten years.
- Force passwords to be at least eight digits.
- Identify unused mailboxes (sometimes used by drug dealers as an untraceable mailbox for transactions).

- Never allow dial tone to be accessible from voice mail.
- Implement a class-of-service program that allows employees or on-premise contractors to have only the features they need. For example, the ability to modify someone else's telephone features is obviously powerful and dangerous if misused — a hacker that gains access to a phone with that class-of-service level could significantly disrupt operations. Review class of service annually.

USING SECURITY TOOLS TO OFFER MORE SERVICES

Although the discussion of security to this point has been from a defensive perspective, there are a few operational enhancements that come out of a good security system. Some of these are listed below.

Use of Voice Verification to Allow DISA

By enrolling employees that normally use calling cards for business (sales people, traveling professionals, etc.) in a voice print authorization system, calling card costs can be significantly reduced. By using an 800 number to call in to the PBX and allowing DISA for an outgoing call (after verification), travelers can obtain the same services at a cheaper rate. Although they would pay for the call two ways (into the PBX and out to another location), the cost of calling card calls is usually so high that the organization still reduces costs. In particular, the cost of calling card international calls and intra-LATTA calls are often well above 800 number rates. Exhibit 11-4 shows a payback analysis using fictitious but typical calling card and 800 number rates. Savings in calling cards alone can pay for the security device because the payback is shown in less than one year. Of course, the payback calculation will vary considerably depending on the number of calls via calling cards, the percentage of users that would be willing to go through the voice registration process, per minute costs of long-distance and calling card usage, and the cost of the verification equipment itself (e.g., Veritel).

Access Voice Mail in Areas of the World Without Touch-Tone Telephones

Using voice-activated-only voice mail (with appropriate speaker voice recognition) allows rotary users to go through menus within voice mail.

Access Special/Confidential Services

For example, Parlance Corporation has a service called "Employee Connector" that allows an individual to have listed multiple phone, pager, cellular, etc. numbers. These numbers can be dialed by saying, for example, "Ms. Doe's vacation home" or "Mr. Smith's New York office." Having this information would be useful for executives and their administrative assistants, but might be too sensitive for the general employee population. By front-ending this service with a security device, it would be practical to use

Exhibit 11-4. **Analysis of potential savings using voice verification in place of calling cards.**

Step 1: Calculate total costs of calling card usage

Number of calling card calls per month	30,000
Minutes of use per month	150,000
Blended cost per minute (average of inter, intra-state, intralatta)	0.21
Setup cost per calling card call	0.25
Cost per minute of calls to and from business locations ("mixed rate" — dedicated to switched location)	0.09
Cost for minutes	$31,500
Cost for setup	$7,500
Total calling card costs for the organization	$39,000

Step 2: Calculate voice verification costs using same assumptions

Cost of minutes — calling in via 800 number	$13,500
Add same number of minutes calling out	$13,500
Total voice verification costs using secure DISA (in and out)	$27,000
Monthly savings assuming 100 percent usage of voice verification	$12,000
Monthly savings if only 25 percent of calls are via voice verification	$3,000

Step 3: Calculate payback

Representative cost of voice verification equipment	$20,000
Monthly savings (25 percent of calls use voice verification)	$3,000
Payback period in months	**6.7**

Note: All costs above are for illustrative purposes only and do not represent actual prices from any carrier or vendor.

---

it — executives would feel confident that only those with a need to know would have access.

SUMMARY

As long as the communication system is connected to the public switched telephone network, it is vulnerable to attack. Reasonable security steps, including class-of-service policies, routine monitoring of traffic patterns by the organization and common carriers, installation of fraud detection equipment, employee education, periodic reviews by an outside party, and close communication with equipment vendors on security issues, can greatly reduce the frequency and severity of security breaks. As the volume of end points in the entire network grows (both the value and the threat of the network increases as the square of the number of end points), ever greater vigilance is required.

Footnotes

1. Post Telephone & Telegraph (telephone company usually owned by a country's government).
2. *Oregon Certified Public Accountant*, October 1994.
3. Web page from Epigraphx LLC, 965 Terminal Way, San Carlos, CA 94070 (http://www.epigraphx.com/faxhacking.htm).
4. Web page from Epigraphx LLC, 965 Terminal Way, San Carlos, CA 94070 (http://www.epigraphx.com/faxhacking.htm).
5. Time Division Multiple Access. Older digital cellular technology.
6. Call Division Multiple Access. This is also called spread spectrum (invented in the early 1940s by actress Hedy Lamarr). It is a highly efficient method of using airspace and has the added benefit of being exceedingly difficult to unscramble.

Chapter 12
Negotiating Carrier Rates and Services

NEGOTIATING OPTIMUM CARRIER RATES AND SERVICES IS TIME-CONSUMING, TEDIOUS, BUT ESSENTIAL. In an era of falling prices, what might appear to be a highly competitive offer may in fact be mediocre and, by the end of the contract, onerous. Both large and small organizations should examine carrier agreements in detail. Because the volume and scope of services used varies so much by organization, carriers cannot really offer a generic good deal. For example, a carrier might offer very low T1 rates with no installation charge, but the organization is a heavy user of calling cards, which the carrier has not priced competitively. With respect to tariff agreements, an organization that does not focus on the details will almost certainly incur excess costs.

GETTING STARTED: COLLECTING DATA ON THE CURRENT ENVIRONMENT

Before an organization can go out for bid (or restructure an existing contract), the following key information must be collected in order to give the carriers sufficient information to make their best offer (if they are so inclined).

- What is the annual volume of minutes?
- Where are those minutes consumed: office or plant locations, type of connection (see later explanation of on-net versus off-net), and domestic or international?
- What method is used for calling: calling card, cellular phone, 800 number (or other toll-free calls)? The breakdown of volumes by service is critical because it significantly affects the weighted average of total costs.
- What data facilities are used: private line (T1 or E1), Frame Relay, ATM, SONET, fractional T1, analog lines, or intrastate digital services?
- Do employees work from home? If so, what volume of minutes or data traffic is incurred?
- What is the architecture of the current network? For example, are remote sites linked to a headquarters office via hub-and-spoke architecture?

- What are the business plans to expand or close offices or plants? Sometimes, carriers will obtain a good deal by specifying a volume of minutes for a particular international location (e.g., United States to United Kingdom traffic). If those volumes are not obtained, penalties are sometimes specified.
- What carriers are used now? What penalties would be incurred if some of the existing locations change carriers?
- Have all locations using the same carrier been consolidated? In decentralized organizations, offices sometimes sign agreements with the same vendor but not under the corporate rate plan. For example, office A may have an agreement with MCI and office B in another state may also have an agreement with MCI, but MCI is unaware that offices A and B are part of the same organization and does not give them the corporate rate. Instead, each office gets the "mom-and-pop" rate typically given to small-volume businesses. In smaller locations, the office manager is sometimes not cognizant of the discounts available from an enterprise-negotiated volume agreement. They know, for example, that the organization uses AT&T as their preferred carrier and assume that the local AT&T office will put them on the correct rate plan. In general, this does not happen. The organization must make the effort to notify its carrier of all offices or plants that should be included in the overall volume agreement.
- What are the long-term trends for the organization? If volumes are expected to change significantly during the duration of the contract, the mechanism for a sliding scale of discounts should be developed.

In order to interpret the agreements, the terms "on-net" and "off-net" (sometimes called "dedicated" and "switched") should be clearly understood. A call that is placed on-net goes through the PBX and then is transferred via dedicated T1 (or E1) access lines to the nearest carrier POP (point of presence). From there, it travels to another location. If this destination location is also contractually part of the same virtual network and has an access T1 or E1 to the nearest POP of same carrier, then the call is completely on-net. The consequence of this arrangement is that in neither case has the LEC (local exchange company) forwarded the call to the long-distance carrier's POP. Because the LEC did not forward the call, it cannot charge the long-distance carrier (sometimes called IXC, for interexchange carrier) for the call. Hence, on-net calls are significantly less expensive than off-net calls. If calls are made from a location that does not have dedicated T1 or E1 lines to the IXC POP (i.e., a small-volume location), then the LEC forwards the call to the IXC POP and charges for that service. Such calls are charged at off-net rates. Mixed rates result from an on-net location calling an off-net location. Typically, an on-net to on-net call will be one third the cost of an off-net to off-net call.

GETTING THE BEST DEAL: A NEGOTIATING CHECKLIST

As mentioned earlier, a generic good deal does not exist. For example, if an organization has only a few Frame Relay circuits, a low price on Frame Relay has little value. The key to optimizing the value of an agreement is to obtain the best mix of volumes, prices, and services most appropriate to the organization's specific needs. The constraints are the commitment/penalties required by the carrier to justify the reduced pricing. Following is a checklist of negotiating points and potential concerns when entering into an agreement with a communications carrier.

- Obtain all proposed prices from the carrier. This may seem obvious; but in some cases, the customer never actually sees all the prices. For example, detailed intralatta prices may be on file as a tariff but never printed out anywhere. The sales representative may say that it would take a dumptruck to print out the entire tariff. Nevertheless, they should be obtainable (preferably in electronic format) so that a spot check (random sample) of agreed-upon rates versus actual billing rates can be run periodically.
- Develop a detailed matrix of all penalties. Typically, a carrier will specify a minimum volume or revenue commitment per year for the duration of the contract. If the contract is voluntarily terminated, the organization will owe the monthly remaining minimum commitments until the end of the contract. Other possible early-termination penalties include return of sign-on bonus and fees for data circuits for which installation charges were waived.
- Understand how discounts are computed. In order to provide the greatest incentives for business, carriers may set up a sliding scale of discounts. For example, domestic dedicated minutes might receive a 25 percent discount for 2 million minutes per year, 30 percent for 3 million minutes, and 35 percent for 4 million minutes. What happens after 4 million minutes? Does the discount drop back down to zero? The organization should always allow for the possibility of a drastic increase or decrease of business (either through growth, acquisition, or divestiture).
- Negotiate for billing at the six-second (or preferably one-second) level. Some carriers will round up to the nearest 18 seconds or even minute. In some situations, such as short polling bursts for remote telemetry units, this could have a substantial impact.
- Carefully define carrier duties. There are some ancillary duties, such as creating circuit diagrams every month or quarter, that assist overall management of the enterprise. Other responsibility questions include: how many on-site representatives will the carrier supply? How often will network optimization be performed (e.g., to find duplicated lines or lines not used at all)? What information will the carrier

193

provide to facilitate internal billing for the organization (e.g., calling card chargebacks)?

- Look at network monitoring provisions. Can the carrier provide response time details? How quickly can they notify the organization that a line is down? What uptime provisions are included, and what credits are provided for downtime?
- Review toll fraud provisions (indemnity). What sites are covered by toll fraud insurance? What procedures does the carrier require to be in place in order for the toll fraud insurance to be in effect?
- How easily can facilities (especially data circuits) be upgraded or downgraded? For example, assume that a 128Kbps Frame Relay circuit is installed and soon afterwards it is determined that a 256Kbps capacity is needed. Can the agreement be amended without penalty? Can the pipe be sized up or down, depending on business needs?
- Take a long-term look at what users are required to learn in order to get discounts. Do users need to dial a very long string of digits to access a special virtual private network that gives the organization better rates? Given that users have real work to do rather than focusing on the mechanics of using the network, should not the carrier offer services that are simple to use? Carriers should be pressured to make all their products as simple to use as possible; users will not go to night school to learn how to use a calling card.
- Compare prices. There are several ways to approach rate comparison: (1) hire a consultant; (2) simply call other organizations to see if the prices are in the ballpark; and (3) consider using some of the packages on the market that maintain current tariffs. The user can both design and evaluate costs for a wide variety of domestic and international networks, including ATM, Frame Relay, TDM, multi-service switches, X.25, and SNA.
- Carefully consider the duration of the contract. In a time of falling telecommunications prices, an organization can get caught with uncompetitive pricing at the end of the contract. Agreements beyond three years are risky.
- Examine the ramp-up period. If network traffic is from multiple business units (belonging to the enterprise) and perhaps multiple vendors, the new carrier may offer a ramp-up period during which volumes can be cutover from previous contracts to the new one, thus satisfying the minimum volume requirements. During this period, the organization gets the full negotiated discounts but is not required to meet the minimum volumes of traffic. There are two constraints on ramp-up timing: (1) the duration is not long enough, there may not be enough time to consolidate the necessary volume of business to satisfy the new agreement; and (2) if the duration is too long, the contract will be extended longer than anticipated (because the contract starts at the end, not the beginning, of the grace period). Because prices are generally falling in

telecommunications, contracts should be negotiated to be as short as possible — while still offering the desired discounts.

- Ensure that a renegotiation clause is included in the contract. Requiring both customer and carrier to review the market competitiveness of the contract on an annual basis can be a win–win for both. The carrier can lower rates in return for an extension of the contract or some other concession. The organization gets lower rates and is able to continue its growth with the same carrier. Unfortunately, when renegotiating an existing contract, it is sometimes difficult to determine whether the offer on the table is competitive. If an organization has an existing contract with heavy penalties for early termination, other carriers are reluctant to go through the expensive process of bidding unless there is a reasonable probability that they will win some or all of the business. Why should they serve merely as a foil to pressure the incumbent carrier to lower prices? Consultants are available that can review proposed rates for reasonableness. However, the acid test for lowest rates is always a head-to-head competitive bid. Therefore, a renegotiation with the existing carrier will always carry some risk of less-than-best pricing. This should be weighed against the costs of (1) extensive negotiating, in terms of organizational manpower, (2) penalties, and (3) the considerable cost of physically changing out access lines, T1s, documentation, contact lists, etc. if a carrier change is implemented. In telecommunications, as for other spheres of life, there is no free lunch.
- Review commitments carefully. Carriers often put complex commitments into long-term agreements (particularly if the agreement is comprehensive and intended to include all or most of the enterprise). For example, there could be a clause that requires 500,000 minutes of international use per year in addition to the overall commitment of 15,000,000 total minutes per year. Sometimes, minimums apply to a specific country. Without a detailed knowledge of where the organization is growing and plans for expansion or reduction, commitments become risky. Punitive clauses are occasionally buried in the fine print. For example, going below the minimum in a single country might trigger a 5 percent penalty on the entire international commitment. If the worst does happen, carriers will sometimes forgive a penalty in order to preserve good customer relations — particularly if long-term prospects for additional sales to the customer are good. Losing a large account can sometimes literally cost an account executive his or her job.
- Determine if combining voice, data, and video requirements into one contract can reduce overall costs.
- Ensure that work-at-home users can access lower-cost network facilities. For example, those users should be able to dial a toll-free number to get access to the organization's best switched rate plan. This is especially important for high-volume users.

- Add cellular long-distance minutes to the total volumes of minutes in order to increase leverage.
- Look at the carrier's willingness to educate the end user. Will an internal Web page be provided for users that instructs them on use of the calling card (e.g., pressing "#" after making a calling card call allows another number to be dialed without repeating the card number).
- Most large carriers offer many ancillary services, such as network-based voice mail, fax-on-demand, fax in the sky, reach me anywhere numbers, and network voice response services. These services should receive some level of discount, even if current usage is minimal.

A COMPARISON SPREADSHEET

Exhibit 12-1 shows the format of a comparison sheet that can be used to standardize the presentation of voice rates from different carriers. If the organization is going out to bid, the format of the responses should be specified in advance so that tabulation is simplified.

If there are specific countries or intralatta traffic that are significant, those locations should be included in the comparison sheet. Another benefit of a structured approach is that *pro forma* calculations can be easily performed; for example, what if the traffic to France doubles or the minutes to Mexico decrease by 30 percent?

OUTSOURCED SERVICES

Many organizations will have a mix of responsibilities: the carrier provides the transport for data, voice, and video traffic, and internal functions (such as wiring) are performed by the internal staff. In some cases, however, entire functions are outsourced. Examples are listed below.

- Videoconferencing
  - facilities, scheduling (global and domestic), and monitoring bundled into one package
  - end-to-end management of all equipment, including codex, internal wiring, cameras, etc.
  - design and engineering of videoconference facilities
- Operations management
  - staffing
  - development of procedures
  - document expectations of users
  - perform user satisfaction surveys
- Network planning
  - tune network performance
  - develop disaster recovery capabilities
  - develop alternate routing and priority management in the event of an emergency

Exhibit 12-1. Example carrier rate analysis.

	Minutes	Carrier A		Carrier B		Net Difference
		Net Cost per Minute ($)	Net Extended Cost ($)	Net Cost per Minute ($)	Net Extended Cost ($)	Carrier A – Carrier B ($)
Intralatta						
Sch A	33,014	0.0905	2,988	0.0988	3,262	(274)
Sch B (Sw→Ded)	1,000	0.0702	70	0.0845	85	(14)
Sch B (Ded→Sw)	45,000	0.0687	3,090	0.0713	3,209	(119)
Sch C	150	0.0356	5	0.0465	7	(2)
Subtotal	79,164	0.0777	6,153	0.0829	6,562	(409)
Intrastate						
Sch A	107,365	0.1020	10,951	0.1120	12,025	(1,074)
Sch B (Sw→Ded)	25,000	0.0666	1,665	0.0678	1,695	(30)
Sch B (Ded→Sw)	310,258	0.0666	20,663	0.0678	21,035	(372)
Sch C	8,014	0.0302	242	0.0412	330	(88)
Subtotal	450,637	0.0744	33,521	0.0779	35,086	(1,564)
Interstate						
Sch A	300,000	0.1179	35,370	0.1245	37,350	(1,980)
Sch B (Sw→Ded)	75,000	0.0800	6,000	0.0884	6,630	(630)
Sch B (Ded→Sw)	1,000,000	0.0800	80,000	0.0844	84,400	(4,400)
Sch C	80,000	0.0601	4,808	0.0689	5,512	(704)
Subtotal	1,455,000	0.0867	126,178	0.0920	133,892	(7,714)
Canada						
Switched	714	0.9987	713	0.9990	713	(0)
Dedicated	24,287	0.3210	7,796	0.3456	8,393	(597)
Subtotal	25,001	0.3404	8,509	0.3643	9,107	(598)
Mexico						
Switched	548	0.3001	164	0.3300	181	(16)
Dedicated	3,419	0.2015	689	0.1987	679	10
Subtotal	3,966	0.2151	853	0.2168	860	(7)
Composite of other international locations						
Switched	15,882	0.6044	9,599	0.6458	10,257	(658)
Dedicated	118,297	0.7012	82,950	0.4580	54,180	28,770
Subtotal	134,179	0.6897	92,549	0.4802	64,437	28,112

Exhibit 12-1 (Continued). Example carrier rate analysis.

	Minutes	Carrier A Net Cost per Minute ($)	Carrier A Net Extended Cost ($)	Carrier B Net Cost per Minute ($)	Carrier B Net Extended Cost ($)	Net Difference Carrier A – Carrier B ($)
0+ Card Service						
Intralatta	11,201	0.3210	3,596	0.3300	3,696	(101)
Intrastate	44,004	0.1478	6,504	0.1544	6,794	(290)
Interstate	89,145	0.2894	25,799	0.2748	24,497	1,302
PR/US V.I.	4,500	0.3699	1,664	0.4015	1,807	(142)
Canada	5,489	0.5124	2,813	0.6154	3,378	(565)
Mexico	458	1.9870	910	2.0110	921	(11)
Overseas	1,248	1.7840	2,226	1.8800	2,346	(120)
Total 0+ Card	106,425	0.4088	43,511	0.4082	43,440	72
800 Service Summary Intralatta						
Switched	45,855	0.2450	11,234	0.2666	12,225	(990)
Dedicated	246,874	0.0578	14,269	0.0601	14,847	(578)
Subtotal	292,729	0.0871	25,504	0.0925	27,072	(1,568)
Intrastate						
Switched	22,154	0.1245	2,758	0.0131	290	2,468
Dedicated	105,478	0.0777	8,196	0.0898	9,472	(1,276)
Subtotal	127,632	0.0858	10,954	0.0891	9,762	1,191
Interstate						
Switched	784,631	0.1245	97,687	0.1302	102,159	(4,472)
Dedicated	457,861	0.0789	36,116	0.0699	31,991	4,125
Subtotal	1,242,492	0.1077	133,803	0.1080	134,150	(347)

Grand Total Difference   34,337
(Cost of carrier A exceeds cost of carrier B)

Note: Above minutes and rates are fictitious and do not represent any actual carrier rates.

Schedule A = switched to switched; Schedule B = switched to dedicated or vice versa; and Schedule C = dedicated to dedicated rates.

— perform statistical analysis on usage and implement capacity planning studies (e.g., establish trunking levels based on Erlanger tables)

When services are completely outsourced, the organization has a tendency to "throw the function over the fence;" that is, the carrier has complete control and the customer merely pays the bill. In such situations, there may not be anyone on the organization's staff that can evaluate performance on a detailed level. When negotiating large contracts for end-to-end services, it may be advisable to use consultants with subject matter expertise to ensure the organization's interests are fairly represented.

MONITORING CARRIER SERVICE LEVELS

Depending on the organization's type of business, service levels can be as important as price. For critical circuits, a carrier should display a sense of urgency. Hospitals, commodity traders, emergency organizations, and others have absolute uptime requirements that should be documented upfront. In addition, noncritical service levels, such as the due date of routine management reports, should be addressed.

Performance measures are not generic. Configurations such as router hardware/software, dial-backup options, configuration, and transport service design vary by organization.

Key points to consider when negotiating carrier service levels include:

- What is the overall network availability?
- What is the mean time to respond and repair circuits and equipment?
- Will the carrier constantly monitor the circuits and notify the appropriate party when it goes down? (Some carriers will charge an additional fee for continuous monitoring.)
- If the same carrier provides both primary and backup circuits, are completely independent, duplicate paths provided? This includes local POPs as well as the IXC portion of the circuit.
- What ancillary services are provided and what criteria are applied to their delivery? For example, if the carrier provides a calling card to the organization's employees, how much time can elapse from the time of card request to receipt by the employee? If internal billing information related to calling cards is to be provided, who does the clerical work of associating a calling card number to an internal department/division accounting code?
- Does the organization maintain its own network monitoring systems (such as Boole & Babbage's COMMAND/Post®)? What is the customer's role versus the carrier's responsibility?
- How does the Help Desk function work? Is the Help Desk 7×24? Can customers, for example, call in trouble tickets and then browse them on the Internet, using a secure ID and password? AT&T, for example,

allows customers to view trouble tickets on the Web and monitor the progress of resolution using their Interactive Advantage integrated platform. Escalation levels should be defined and unresolved problems should be "aged."

- What problem reporting management reports are available? As a minimum, the reports should detail total downtime, number of trouble tickets, explanation of problems (whether carrier, customer, other, or unknown cause), time to resolve, and circuits affected.

- How does the carrier monitor and report toll fraud attempts/activities? Carriers tend to concentrate their toll fraud monitoring facilities in large network command centers. The procedures for customer contact are relatively rigid and must be predefined. They use sophisticated algorithms to detect fraud, but need up-to-date contact lists to ensure timely response to individuals in the organization that can make decisions. Organizations that have a lot of calling card usage are constantly faced with the problem of notification. If executives going out of the country give their calling card number out to someone else in the organization and both use the card within a few hours of each other, it looks like calling card fraud to the carrier. In that case, who gets called to make the decision whether to cancel the card or continue to allow calls to go through? In a large organization, calling card services (getting to the right person, distributing cards, etc.) can easily be a half- or full-time job. Service levels should state who is responsible.

- Are end-to-end responsibilities defined? A worst-case scenario is for a complex outage to occur with an organization's carrier, claiming that their network is fine but the customer-provided equipment is malfunctioning or perhaps the local exchange carrier's equipment is bad. Agreements should clearly state where responsibility starts and stops. Also, as a practical matter, carrier technical personnel should be willing to work with the customer on a problem, even if the problem appears to belong to someone else. For example, after testing all the circuits, in-house wiring could be the culprit for a down line. The carrier may charge on an hourly basis for the assistance, but should not refuse to get involved (for a critical function).

- What is the service level for technical field support? Two- or four-hour response time after business hours? Response times for remote areas should be specifically addressed.

- Is on-call support consolidated into one 800 number? Or is there a different number for each service?

- Does network monitoring include troubleshooting of protocol errors, analysis of recurring errors, network response times, proactive monitoring of congestion, feature availability, utilization, customer satisfaction, and access violations?

- Include miscellaneous services such as audio- and videoconferencing. As with all services, internal billing issues should be considered. For example, assume all audioconferencing bills are lumped together in a single bill to XYZ Corp. Unless service levels are defined otherwise, it will be the responsibility of XYZ Corp. to spread those costs to individual departments, do research on exceptions, etc.

EXAMPLE OF CARRIER SERVICE-LEVEL SPECIFICATIONS

Following is an example service-level specification from AT&T. Service levels will of course vary by carrier, services purchased, and other criteria, but the information below should give the reader a rough estimate of what can be obtained in the market.

Frame Relay (from AT&T)[1]

Provisioning: If an agreed-on due date is missed for a port or PVC, recurring charges on the port or PVC are free for one month.

Restoration time: if a customer reports a Frame Relay service outage (even if the problem is with local access) and it is not restored in four hours, recurring charges for the affected ports and PVCs are free for one month.

Latency: If the customer reports a one-way delay from service interface to service interface (SI-to-SI) across the Frame Relay network of more than 60 milliseconds and AT&T cannot fix the problem in 30 days, recurring charges for the affected PVC are free each month until repaired.

Throughput: If 99.99 percent of the packets offered to the Frame Relay network within a PVC CIR (committed information rate) are not successfully transported through the network and AT&T cannot fix the problem in 30 days, recurring charges on the PVC are free each month until repaired.

Network availability: If the customer's network is not available at least 99.99 percent of the time each month, the customer receives credits commensurate with network size.

MAINTAINING OPTIMUM DISCOUNTS IN A DECENTRALIZED ORGANIZATION

From the perspective of the chief accounting officer, the bottom line of any telecommunications agreement is the cost of services *before* the agreement, minus the cost of those same services *after* the agreement. Some particular costs (e.g., audioconferencing) may not go down at all in the new agreement — it is the *total* that counts.

Unfortunately, in a decentralized environment, there is a tendency to "cherry-pick" the contract. Field offices, far removed from the comprehensive analysis done to get the best overall pricing, see that some other carrier can give a better deal on a single circuit or limited service (e.g., Billy

Bob's Frame Relay resold from a major carrier). By selecting the cheapest offering at every location, the organization is exposed to the following.

- Volumes could dip below minimum requirements, initiating penalty clauses.
- Introduction of a plethora of vendors may cause additional work in administering the complete network.
- With many contracts starting and stopping at different times, it is more difficult to move the enterprise to a new, more economical agreement.
- The billing workload can increase due to a larger number of vendors.
- Many times, exceptional rates are loss leaders that will not be repeated for all new business from the same carrier.

With many vendors, it is difficult to obtain reliable statistics on overall usage. This not only affects negotiations when trying to consolidate volumes for best pricing; it also makes network planning difficult. For example, assume that multiple PBXs and servers are to be linked by Cisco Stratacom/3810 ATM units using T1s to various sites. Trying to determine what T1s go where and which ones can be consolidated is more difficult with multiple vendors/resellers.

Of course, there are reasons to have multiple carriers. For example, if a link between two locations is absolutely critical, a second carrier can provide excellent redundancy — provided there are no agreements between the two carriers to share physical facilities at any point along the path. In this case, there is a justifiable business reason for doing so, not simply because a single circuit looks like a great deal.

SERVICE LEVELS AND ORGANIZATIONAL REQUIREMENTS

A well-executed service-level agreement serves to codify what is critical for the enterprise. If cost is the driver, then service requirements may be more lax. If uptime is an absolute, the lowest-cost carrier may not be an option. Larger organizations may have units with differing objectives. Hammering out service-level agreements on the front end may help pinpoint real requirements that are essential for success. Everyone wants zero downtime and 15-minute response. Are they willing to pay for it?

Service levels are more than simply how fast something gets fixed. Some facets of service levels address intermediate steps such as:

- How quickly will there be an initial response from the Help Desk?
- At what point does level 2 escalation occur and who is notified?
- How is the trouble tracked?
- Who will coordinate with all vendors and organizational representatives involved?

SUMMARY

There are significant benefits to careful monitoring and negotiating carrier rates. After the current volume data is gathered, the organization can provide carriers with aggregate volumes to obtain the lowest rates. The pure financials must be tempered by service-level agreements to ensure that the quality of services are adequate.

Notes

1. AT&T Web page, AT&T Frame Relay Service — Domestic SLAs, http://www.att.com/press/ 0198/980127.bsc.html, January 1998.

Chapter 13
Outsourcing Options and Service-Level Agreements

FOR BETTER OR WORSE, TELECOMMUNICATIONS OUTSOURCING DECISIONS affect an organization's level of service, costs, and innovation for years. Few business trends have had as much starkly contradictory press. For example, outsourcing has been praised as delivering:

- lower costs
- faster ramp-up time for projects
- higher levels of expertise
- enhanced career paths for technically oriented individuals who have "no place to go" in an organization whose business mission has nothing to do with telecommunications

Alternatively, outsourcing has been vilified as the agent of:

- higher costs
- resistance to change (maintain status quo rather than innovate)
- inadequately trained technical staff (training cuts the service provider's bottom line)
- indifference to the customer's culture (service provider has inflexible procedures and is not responsive to a dynamic, rapidly changing business environment)

Like any good partnership, the success of outsourcing depends on the culture and technology fit between service provider and customer. The following chapter sections outline the major factors that should be considered when entering into or renegotiating a telecommunications outsourcing agreement.

THE IDEAL CASE

A few years ago, EDS, a large service provider based in Plano, TX, developed a concept video for its customers that illustrated outsourcing at its best. A music conductor wanted to have a violin sonata played simultaneously around the world, with locations in the United States, Japan, and other countries. EDS provided all the communications technology from A to Z. The conductor, as a user, knew nothing about the technology, but wanted perfect, simultaneous communications. EDS delivered the infrastructure for the live, around-the-world concert and everyone was happy.

Unintentionally, however, the video revealed the environment that creates maximum satisfaction for an outsourcing engagement:

- The customer knows or cares little about the underlying technology (e.g., the conductor did not argue that costs could be reduced by using multiple ISDN lines rather than a dedicated T1).
- The customer wants the service provider to "handle it" — take care of all details and merely present the desired business result.
- The project has a beginning and an end, with a cap on expenses ("not to exceed").

Other business conditions that mesh well with outsourcing include:

- a relatively stable business that generates many transactions or activities that can be quantified and costed
- a special project that requires narrow expertise and must be done quickly
- a small business that could not afford to hire in-house staff for telecommunications
- the need to make costs more predictable
- the flexibility of pricing that allows a cash-strapped company to pay less for a given bundle of services on the front end of a contract and relatively more for the same bundle of services at the end of the term
- elimination of an internal group that has not been responsive to end users or has been profligate with expenses

CAVEAT EMPTOR: THE DOWNSIDE

Harold Geneen once remarked that no single business theory can address all the complexities of the business environment. So, it is with outsourcing — it has limitations that make it effective for some organizations and unsuitable for others. Following are some of the potential problems that could be encountered in outsourcing arrangements.

- The service provider, no matter how honest and sincere, has a different bottom line than the customer. For example, in practice, it is most difficult for a service provider to tell a client that an expensive, proprietary system should be scrapped — if it means a significant reduction in revenue to the service provider.
- Service providers are not uniform in their capabilities. Simply because outsourcer A has 50,000 employees and clients worldwide does not mean that it has expertise in all areas. Some service providers include

communications as an ancillary service to round out a full-service information technology/communications offering. For example, their communications expertise may not be on par with their database expertise.

- Some outsource firms may not have the domestic or international coverage required. These firms may have effective agreements with other service providers to ensure adequate maintenance coverage. NEC partners, for example, rely on a network of providers to cover domestic and international markets. Depending on the criticality of the service, such arrangements may or may not be effective. Having a single, unified maintenance organization can provide a stronger level of accountability. The vendor's number of "feet on the street" should be considered when making an outsource decision that extends to multiple locations.

- Service providers can become complacent regarding technical training for their employees. In long-term arrangements, who pays for training must be specified; otherwise, it is tempting for the outsourcer to use OJT rather than formal technical training to keep their employees up-to-date.

- The major service provider may attempt to become the sole front end for all vendors providing services to the client. Thus, a client's management may not hear about a promising technology that could significantly reduce manpower requirements to operate communications technology. As an example, assume a customer has an older PBX that is nodal in architecture, meaning that when a user moves from one location using node A to another location using node B, significant wiring changes on the frame are required. A newer PBX, such as the Lucent Definity GR3, is not nodal and requires no wiring changes for moves between already hot ports (every port is equally accessible). Such technology improvements can significantly reduce manpower requirements for moves, adds, and changes (MACs). The service provider has little incentive to bring these ideas to the attention of customer management — *unless* some mutual savings participation has been built into the contract.

The above comments should not be interpreted as a pejorative orientation toward outsourcing firms. If all the issues and economic factors are fully understood and written down by both parties, outsourcing can be highly effective in achieving service and financial goals. The following section outlines key ingredients in a sound and effective agreement between the two parties.

NEGOTIATING THE AGREEMENT

The key principle is: *do not assume anything*. Everything than can possibly be conceived (especially failure points) should be addressed clearly, in writing, and with examples if necessary. In addition, the potential client should retain the services of an outsource consultant if the agreement is of long duration and a large dollar amount. The service provider, of necessity, knows significantly more about personnel costs, equipment costs, service requirements, long-term industry trends, etc. than does the prospective client. Also, outsourcing agreements are sometimes reached in times of organizational stress and important details are unintentionally omitted. Outsource consultants should be involved in big deals — for the long-term benefit of both parties. Someone who has gone through the process many times can bring to the table key issues that would not otherwise be addressed.

Following is a partial list of factors that should be addressed in any major outsourcing agreement (a comprehensive list is not possible because each agreement must include unique business needs).

- What is the duration and scope of the agreement? What locations, equipment, users, vendors, etc., are to be included? What services are to be provided?
- Under what conditions can the agreement be renegotiated? How are specific requirements to be enforced?
- Are all costs listed? Commodity prices, such as cost per minute to specific locations or at least the surcharge above carrier rates should be stated. Hourly rates, on-call, rush, after-hours, holidays, and other manpower costs must be included. Administrative charges for ordering equipment should be explicit (e.g., 2 percent handling fee).
- Are levels of expertise described, with specific monthly prices? A switch technician should be billed at a different rate than a MAC technician.
- What is coverage and billing for holidays, sick, vacation time, and training time? Who bears the cost?
- Who pays out-of-pocket expenses and temporary relocation for short duration projects?
- Under what conditions can the agreement be terminated?
- Who pays for incidental work tools, such as PCs, printers, etc.?
- Is the agreement exclusive? If so, how does the service provider bill when another vendor must be brought in to do work that the service provider cannot perform (due to lack of expertise or insufficient manpower)?
- Under what conditions do fees change? Is there an annual competitive review to ensure that the service provider is charging rates that are not significantly higher than going market rates?
- Do clients have full audit rights to see all their data, along with the source information that is used to generate the bill from the service provider?
- What are the penalties for failure to perform? Are they sufficiently well-defined to be enforceable?

- Who controls the technical architecture? For example, during the early days of transition from mainframe to client/server technology, some service providers were extremely resistant to change — understandably, because their workforce was largely trained in traditional third generation languages such as COBOL and retooling was a significant expense. The author strongly recommends that the client retain control/direction of the technical architecture.
- What is the culture of the service provider? Can end users directly contact the provider's "worker bees," or is it necessary to go through a front line manager? Is the work model hierarchical or distributed/decentralized?
- Under what conditions can service provider personnel be hired by the client? What are the fees should any such transfers occur?
- To what extent will the outsourcer provide internal billing services? For example, will their electronic invoices carry accounting coding that allows the charges to go directly into the client's general ledger system? Will they segregate long-distance bills by responsibility codes?
- Are service levels defined in sufficient detail to allow proper monitoring and enforcement of agreements?

The last bullet above is key to making an outsource agreement work (regardless if it is between two independent firms or internally between one department and another). The section below describes a typical service-level agreement for telephony services.

SERVICE-LEVEL AGREEMENTS (SLAs)

Judging the level of service provided by a vendor is somewhat like asking the question, "Are airline seats too small?" The answer depends on the criteria — for a child, they may be too big. By writing down exact expectations, both parties develop a scorecard that targets performance toward desired goals. It also serves as a vehicle in the beginning of the agreement to understand what is important to the business. For some organizations, a four-hour response time after 6 p.m. is acceptable; for a 24-hour a day electricity trading firm, five minutes may be the maximum response time allowed.

Typically, service-level agreements require considerable effort to develop and negotiate. They must be real in the sense that they are tailored to the business and communications environment rather than merely copied from boilerplate agreements. Following is a list of items (not comprehensive) that should be addressed when developing SLAs for communications services.

- General objectives of the SLA include, among other business-specific items: (1) the need for written agreements between the customer and service provider, (2) quantification of deliverables and the ability to measure service quality, (3) facility for conflict resolution, and (4) intent to continuously modify the agreement over time to reflect changes in the business environment.

- Specific objectives examples might include: providing dial tone and phone mail services to end users, with uptime of 99.9997 percent; and maintaining long-distance trunks (via redundant carriers) 99.9997 percent of the time.
- Listing of customers, that is, what offices, buildings, groups of employees, etc., are to be covered, and listing of all vendors to provide services.
- Responsibilities of the customer: specific duties might include: (1) providing a point of contact for the service provider, (2) reporting problems to the service provider's Help Desk, (3) assigning resources for telephony projects that require customer input and guidance, (4) reasonable notification of business changes, such as a subsidiary purchase, that would significantly affect volumes or the architecture of telephony services, (5) prioritizing problems, if required, and (6) paying invoices in a timely manner.
- Responsibilities of the service provider will, of course, vary by the specifics of the agreement. A basic bundle of services might include:
  — installation, tuning, and maintenance of the voice communications system, including PBX, voice mail, communications servers, application servers such as fax-on-demand, IVR, and CTI equipment, and any other adjunct processors
  — administration for specific sites between the hours of 6 a.m. and 7 p.m., Monday through Friday
  — after normal business hours and on weekends and holidays: one hour to respond to a page; two hours to be on-site, if required
  — traffic engineering to ensure trunking is adequate to meet a P .001 (0.1 percent of outgoing calls blocked) standard of blocking
  — installation of circuits, including T1, T3, OC3, etc., as required
  — installation of Centrex service, as required
  — call detail reporting via remote polling
  — end-user training, including the preparation of pamphlets, intra-net materials, etc. Topics should include use of telephone, voice mail, and any special long-distance access codes. Special training must be made available to administrative assistants, Help Desk personnel, and any employees functioning in a call center environment
  — moves, adds, and changes
  — Help Desk services between 7 a.m. and 6 p.m.
  — consultation as required for telephony systems, including CTI applications, telephony package installation, IVR, etc.
  — directory services (online telephone book) for multiple sites. Coordination of AMIS (audio message interchange specification) sites to

    ensure that voice mail messages can be transferred from one voice system to another
- — remote administration (if applicable) for sites outside the normal coverage area
- — maintenance of a domestic and global dialing plan, if required
- Caveats. Services may be interrupted for reasons outside the provider's control: fire, flood, and other natural disasters (*force majeure*). Also, if the common carrier servicing the customer is down then local, long distance, or both services can be affected. Of course, backup circuits, different carriers, etc., can be included as part of the SLA.
- Technical architecture. The service provider can more easily control costs and responsiveness if the technical environment is uniform. For example, if all the PBXs are from Northern Telecom, then spare parts can be consolidated, training in multiple architectures is not required, etc. The architecture and requirements for future purchases of equipment and software should be well-defined.
- Performance measurements. Exception reports and management summaries need to be defined, along with the reporting frequency. Web-based reporting by department is increasingly replacing bulky report mailings to department heads.
- Problem management defines how customers are to report problems, and describes Help Desk personnel, hours, type of software (e.g., Peregrine or Remedy), impact statement, parties affected by the problem, assignment of recovery priorities, estimation of resolution time, *post mortem* of the problem, and other detailed information.
- Priority scales. Priority should be defined, along with actions to be taken for each level. For example, priority 1 problems may require an update every 15 minutes, whereas priority 3 problems may only need daily updates.
- Notification procedures for planned outages and maintenance. The following should be defined:
  - — identification of the individual or business unit is the main point of contact for telephony services
  - — the lead time for outages (e.g., 24 hours)
  - — listing of systems that will be out and parties that will be affected
  - — description of effect on the end user (e.g., lack of dial tone or merely loss of voice mail services)
  - — description of project plan along with backout options
  - — identification of project leader and backup personnel
- Escalation procedures. The sequence of escalation (e.g., from Help Desk to first line manager, to general manager, to vice president) should be defined. Categories of problems include:
  - — normal malfunction of equipment or service
  - — repeat problem
  - — priority (premium) service problem

Time guidelines should be developed for escalation (e.g., call the first line manager after two hours of downtime, general manager after three hours, and vice president after four hours). Multiple points of contact — work and home numbers, cellular numbers, pagers, e-mail ID, etc., need to be regularly updated.

- Cost of service. Charges for services visible to the user are the most logical mechanism for cost recovery (plus profit for an outside provider). Dial tone, station equipment, long-distance services, phone mail, and other visible services should be listed in a table with pricing per user. Prices may vary according to the size of the installation or other factors. Charging for many "under-the-hood" items (such as call accounting software) may be counterproductive because the users are not familiar with the purpose of the hardware/software and are suspicious of possible "padding" on the invoice.

Appendix C shows a sample service-level agreement.

CALL CENTER OUTSOURCING

Call centers will be covered in depth in a later chapter, but there are a few items that should be reviewed from the perspective of outsourcing agreements.

- Right to monitor. Large call centers (as service providers to other firms) operate on relatively thin margins. It is important that the customer retain the right to use remote monitoring software to look into the PBXs, IVRs, and other voice systems operated by the service provider. This provides a level of assurance that calls are not being transferred to a secondary or tertiary provider where call statistics cannot be monitored easily. Of course, this requires software that can separate various customers from each other within the provider's systems.
- Who owns the minutes. Assume, for example, that a large toy company introduces a major ad campaign for its jabberwocky vorpal sword. The company publishes an 800 number and the response is ten times the anticipated volume. To service its customers, the company outsources the order fulfillment to a call center service provider. The 800 number receiving all the calls is redirected from the company's premises to the call center service provider. Furthermore, the call center agreement specifies that the call center owns the 800 number. Now all those minutes, instead of increasing the volumes of the toy manufacturer (and allowing greater volume discounts), belong to the call center. The customer may or may not get a best per-minute price. The unit charge per order fulfillment has call center services and long-distance charges bundled into a single price — making it difficult to evaluate the economics of the agreement. In general, the customer should retain ownership of any 800 numbers that are used for large volumes of calls.

An additional benefit to the customer is that should services from the call center prove unsatisfactory, the 800 number can be terminated at another call center at the customer's request.

SUMMARY

Outsourcing agreements, if carefully constructed and closely monitored, can provide the right telephony services for the right price at the right time. However, to make it successful, service-level agreements and an understanding of outsourcing processes must be in place. Finally, the responsibility for monitoring costs and effectiveness can never be abrogated or assigned — constant vigilance is required for success.

Appendix A
Pallas Athene Reproductions, Inc. Request for Proposal: PBX and Voice Mail Systems

Contents

1. INTRODUCTION

1.1. Proposal Request

Pallas Athene, Inc. (hereafter known as "PAR, Inc.") is requesting proposals and quotations in order to select a PBX and voice mail system to serve as its primary voice communications facilities for the Knoxville and Chatanooga manufacturing sites. The successful bidder will provide replacement of the current HSAVM switches (HSAVM models 50 and 70), and provide interfacing capabilities with its networks, backbone, and application servers to support the corporation's CTI and call center activities. PAR, Inc.'s call centers are managed by various business units throughout the corporation. Voice communications are operated 24 hours per day, 7 days per week. The uptime of the proposed system (both voice mail and PBX systems) is expected to be 99.9999 percent with exception of scheduled maintenance time.[2]

1.2. Business Environment

PAR, Inc. is the world's largest manufacturer and distributor of top-quality classical Greek statuary. With approximately $500 million in assets and sales of over $1 billion per year, PAR, Inc., distributes statuary in 85 countries. Two manufacturing plants, located in Knoxville and Chattanooga, TN, use native Tennessee marble to provide reproductions that closely match the original 5th century masterpieces. Our products appeal to high-income, upscale customers who are extremely conscious of material and artistic quality.

PAR, Inc., operates a 7×24 call center out of the Knoxville, Tennessee office. All orders are shipped within 24 hours of receipt.

PAR, Inc.'s internet address is www.par-aristides.com and its common stock is traded under the ticker symbol, "PARARI."

1.3. PAR, Inc.'s Communications Vision
PAR, Inc.'s customers, employees, suppliers, and other interested parties should have the tools and infrastructure available to communicate effectively and efficiently. Getting the right information to the right person in the most appropriate medium is critical for our success.

Over the next few years, PAR, Inc., will develop a communications and collaboration infrastructure that supports the following goals:

- Reach anyone in the PAR, Inc., community via the most appropriate technology — voice, fax, e-mail
- Support a high-quality collaborative environment, including audioconferencing, PC-based videoconferencing, white boarding, and other communications facilities
- Provide a "universal" directory that allows interested parties to contact the right persons or groups within PAR, Inc., for a specific business need
- Provide the capability to handle large-volume, retail call center functions, including CTI (computer telephony interface) to allow caller/database interface, screen pops, etc.
- Empower employees to more fully use and manage their messages by providing an integrated desktop interface for all message media (voice, fax, e-mail, video)
- Improve PAR, Inc.'s competitive position by capturing relevant customer/marketing information available from the communications infrastructure

2. SCHEDULE OF EVENTS

2.1. Schedule Table
The following schedule highlights the major events and dates of the selection process:

Date	Time	Activity
Wednesday, February 4, 2001	7:00 p.m.	RFP mailed
Friday, February 6, 2001	4:00–7:00 p.m.	PAR, Inc. answers vendor questions
Friday, February 20, 2001	5:00 p.m.	Responses due
Monday, February 23, 2001	8:30 a.m.	Vendor #1 discussion

Date	Time	Activity
Wednesday, February 25, 2001	8:30 a.m.	Vendor #2 discussion
Friday, February 27, 2001	8:30 a.m.	Vendor #3 discussion
Tuesday, March 3, 2001		Vendor selected and contract negotiation begins
Wednesday, September 3, 2001		Latest installation date

Note: Vendor discussion/presentation will be held for one-half day with each respondent. PAR, Inc. will contact each vendor for scheduling. Additional sessions will not be granted. Questions after 2/6/01 should be e-mailed to PAR, Inc. (attn: Smith@par-aristides.com). All answers and all questions will be e-mailed to all respondents.

3. FORMAT AND TERMS OF THE RESPONSE

3.1. General

PAR, Inc., expects all respondents to provide RFP responses concisely and brief to point. Supplemental detail information should be available upon request. (Please, no "shovelware.")

An executive summary, no longer than five pages and in MS Word, must accompany your submission.

Respondents will submit twelve (12) copies of their proposal. PAR, Inc., prefers electronic format (MS Word, Excel, Power Point, etc.) but recognizes that some material may not be available electronically. PAR, Inc., will not reimburse respondents for any cost associated with the preparation of a response to this RFP or any subsequent presentations or negotiation sessions conducted prior to a contract award. Conciseness and brevity are encouraged.

Where practicable, respondents should write answers directly below questions (in electronic form). Where the RFP merely specifies PAR, Inc.'s intent or requirement, please mark "agree" or "do not agree."

3.2. Date and Mailing Address for Submission of Responses

All responses to this RFP must be received by 5:00 p.m. Central Time on Friday, February 20, 2001. Responses should be addressed to:

Alfonzo Smith
PAR, Inc.
7901 Aristides Street, Suite 5BC
Knoxville, TN 37938

All responses received after the time specified will not be considered. Those responses will be marked "LATE" and returned to the respondent unopened.

If responses are sent by mail, overnight delivery service, or by courier, the respondent assumes all risk of late delivery to PAR, Inc.'s evaluation team.

3.3. Questions

If PAR, Inc., so chooses, significant information may be transmitted to all respondents simultaneously via e-mail. Vendors should have all questions ready no later than February 6, 2001 (best to submit in advance but all questions will be answered in the 2/6/01vendors meeting).

3.4. Documentation Requirements of Response

The complete proposal response to the RFP must include:

- Point-by-point response to each item
- <u>Written list of exceptions (if any) to the requirements of all sections</u>
- Cover page for the response signed by an authorized representative of the company

3.5. Specific Products, Quantities, and Prices Required

All respondents must provide a spreadsheet containing specific recommended product and service description, and specifications, list price, and allowable discount percentage for each product. PAR, Inc., requests this information to be filed in an electronic medium format.

3.6. Caveats

PAR, Inc., is committed to fairness and objectivity in its evaluation of products. If this document refers to any specific vendor's product, it is entirely accidental and unintentional.

4. PROPOSAL EVALUATION

4.1. Evaluation Criteria

Fred A. Pericles Business Consulting group and PAR, Inc.'s Switch Evaluation Committee will review responses. Recommendations will be made to PAR, Inc. Management, who will make the final decision. Evaluation criteria include:

- Vendor and vendor's products capacity to match PAR, Inc.'s voice communications requirements
- Ability to deliver the service levels on hardware and software required by PAR, Inc.
- Experience in providing quality services that are similar in content and scope to those requested by PAR, Inc.
- Level of reliability, technical innovation, and "fit" with PAR, Inc.'s business needs

- Customer references
- Cost

5. VENDOR QUALIFICATIONS

5.1. Vendor Agreements

The successful vendor must conform to requirements as stated in PAR, Inc.'s Master Services Agreement.

If your firm is the winning bidder, PAR, Inc.'s expectation is that you will sign the agreement with the text essentially in the format as attached.

6. REQUIREMENTS

6.1. "Most Critical" Requirements

The following requirements must be met for your proposal to be considered. Omission of any of these requirements results in automatic disqualification.

Two switch configuration should be transparent to users. (In other words, users in Knoxville and Chattanooga do not need to know which system they are on and do not change the way they operate their handset based on switch domain.)

6.1.1. Describe the Chattanooga survival plan in event of a main system outage (PAR, Inc. building) including Order Center operations.

6.1.2. Integration with Hoplite log systems

6.1.3. Full virtual office capability (no operator/technician intervention required)

6.1.4. Consideration of space limitation for installation of the proposed main switch (must be able to fit into the existing floor space)

6.1.5. Technical training support for PAR, Inc., voice communications agents and third-party contractors. This includes detailed technical system level training to enable PAR, Inc., if it so chooses, to "self-maintain" its PBX and voice mail systems.

6.1.6. E-mail integration capability (must be current available, not a "future") with Lotus Notes version 5.0 and above, as well as e-mail protocol standards such as browser, POP3, SMTP, LDAP4, etc.

6.2. General Requirements

6.2.1. Cost Effective

The proposed system must be cost effective and demonstrate a reasonable return on investment over a period of time and reasonable protection of investment from obsolescence of technology.

6.2.2. Enhanced User Productivity

The proposed system must be able to readily demonstrate improved user productivity.

6.2.3. Intuitive Operation

The proposed system must be intuitive to operate for all users, attendants, and administrator, and require only a minimal amount of training to perform basic functions.

6.2.4. Compliance with Standards

The proposed system must be able to comply with national and international standards established by recognized bodies. Both hardware and software must be designed and implemented as an open architecture.

6.2.5. Internetworking

The proposed system must be designed such that voice, data, and video applications can be readily switched, routed, and where applicable conferenced between all facilities. Interface to the network will be Centrex, T1, HX64, ISDN, ATM, or Frame Relay. Networks may be a combination of public, private, VAN, proprietary, and bypass.

6.2.6. Management and Administration

The proposed system must have the capability of centralized and decentralized management and administration, as well as the capability for remote alarming of critical events. The proposed system must have effective and automated administration of MAC (move, add, and change).

6.2.7. Modularized Design

The proposed system must be a modularized architecture that allows ease of interface with common hardware. This architecture must be interchangeable between systems and sites if system reconfiguration becomes necessary.

6.2.8. Phased and "Flash cut" approach

Respondent should outline in detail the options for cutover from the existing systems to the new systems. Phased (floor at a time or range of numbers at a time) versus flash cut should be addressed.

6.2.9. Project Management

Timing, personnel involved, project management activities and reporting, and general day-to-day procedures during implementation should be discussed. Include, as part of your response, a high level Gantt chart of expected implementation steps.

Resumes of key personnel implementing and managing the cutover should be included in the response.

- 6.4.3.12.

6.3. Standards
6.3.1. Support of the following standards is required:
- 6.3.1.1. AMIS
- 6.3.1.2. QSIG
- 6.3.1.3. TAPI
- 6.3.1.4. TSAPI
- 6.3.1.5. JTAPI

6.4. System Design and Integration Requirements
6.4.1. The bidder's proposed solution should fit PAR, Inc.'s existing infrastructure

6.4.2. Switch descriptive information
- 6.4.2.1. Brand and identifying information
  - 6.4.2.1.1. Manufacturer:
  - 6.4.2.1.1. Model:
  - 6.4.2.1.1. Registration Number:
  - 6.4.2.1.1. Ringer Equivalent Number (REN)

6.4.3. Switch Architecture
- 6.4.3.1. The proposed Telecommunications Processor must be a digital switching system with integrated voice and data communications.
- 6.4.3.2. Two-switch configuration (should be transparent to users).
- 6.4.3.3. The proposed switch must have stored program control, self-diagnostic routines, modular design, and duplication of critical subsystems.
- 6.4.3.4. Does the proposed switch have any single points of failure that could cause user-affecting downtime? Would a call in process be affected by a single component failure?
- 6.4.3.5. The proposed system must have redundant processor to back up the main processor in event of failure.
- 6.4.3.6. Provide the proposed switch description.
- 6.4.3.7. Describe redundancy of the switching bus if it is optional.
- 6.4.3.8. Describe the time slot architecture of the proposed system.
- 6.4.3.9. Describe time slot utilization for a typical voice call.
- 6.4.3.10. Describe if the proposed system can be configured for non-blocking operation.
- 6.4.3.11. Describe the maximum number of ports that can be supported in a non-blocking configuration.
- 6.4.3.12. Provide the capability of the proposed system to support growth and changes to the system.

6.4.3.13. Provide how load balancing can be handled with growth and changes.

6.4.3.14. Describe the modular architecture of the proposed system.

6.4.3.15. Describe dynamics of system configuration from the minimum to maximum.

6.4.3.16. Describe if key component hardware modules are interchangeable if the two switch architecture is presented.

6.4.3.17. Provide the main memory allocation scheme.

6.4.3.18. Describe how the proposed system can provide for redundancy of system memory.

6.4.4. System Reliability

6.4.4.1. Describe your process ensure reliability of the proposed system.

6.4.4.2. Describe all duplication options available with the proposed system.

6.4.4.3. Chattanooga survival plan in the event of a main system outage (PAR, Inc. building)

6.4.4.4. If the main processor is duplicated, each one must be individually capable of handling the total system traffic load without degradation in service.

6.4.4.5. If the main processor is duplicated, it must operate in a hot standby mode. All memory and databases must be resident in both units.

6.4.4.6. If the main processor is duplicated, the proposed system must be able to switch from one to the other on a scheduled or emergency situation without affecting any switch operations, including established calls.

6.4.4.7. If the main processor fails, what is the impact of the system?

6.4.4.8. Provide the maximum number of lines/trunks impacted if a circuit card fails.

6.4.4.9. State the busy hour call completion and busy hour call attempts capacity of the proposed system.

6.4.5. Networking

6.4.5.1. Provide a full integration capability with PER-FX Order Entry system.

6.4.5.2. Multiple switches must be connected in a private network.

6.4.5.3. When multiple switches are connected in a private network, a high level of feature transparency must be provided.

6.4.5.4. A four or five-digit extension numbering plan must be supported across multiple switches.

6.4.5.5. The switch must support dialing of seven or 10-digit numbers to an interexchange carrier virtual network.

6.4.5.6. The network (originating switch) should be able to provide locally generated tones (e.g., busy signal).

6.4.5.7. The switch must be able to send data calls to one location and voice calls to another location using the same numbering plan.

6.4.5.8. Consistent displays must be provided across the network (connected party name and number and calling party name and number).

6.4.5.9. The proposed vendor should provide integrated voice messaging throughout the network with minimal impact on the grade of service using a combination of private and public trunking facilities.

6.4.5.10. E-mail integration capability (must be current available, not a "future") with Lotus Notes version 5.0 and above, as well as e-mail protocol standards such as browser, POP3, SMTP, LDAP4, etc.

6.4.6. Integration

6.4.6.1. Can all ports be used for call answer, voice mail, outcalling or automated attendant, or do ports need to be dedicated to one or more features?

6.4.6.2. Do ports have to be used for activities like lighting message-waiting lights?

6.4.6.3. Does the system receive call disconnect information from the switch and immediately terminate a session?

6.4.6.4. Is it recommended to end a call answer recording with a touch-tone command or can the caller simply hang up and leave the port immediately available for the next call?

6.4.6.5. Is it recommended to tell the system "good-bye" when a subscriber finishes a voice mail session?

6.4.6.6. If a subscriber retrieves some, but not all, new messages, will the message-waiting light stay lit?

6.4.6.7. If a port has trouble, can the system take it out of service and automatically notify the switch to stop sending calls to that port (without manual intervention)?

6.4.6.8. Does the system provide call coverage for ring/no answer and for busy calls?

6.4.6.9. Does the system provide call coverage for send-all-calls?

6.4.6.10. Are there difference system greetings that play for busy vs. other coverage reasons?

6.4.6.11. If a transfer is initiated to a busy station, or an invalid number, will the system tell the caller and give him another opportunity to transfer?

6.4.6.12. Can the system synchronize its clock with the switch's clock?

6.4.6.13. Can one system support multiple subscribers on remote switches?

6.4.6.14. Does the system provide call coverage for busy calls, for ring/no answer calls, and for send-all-calls?

6.4.6.15. Can one system support multiple subscribers on remote switches? If so, can subscribers on their remote switches have voice mail in their coverage paths?

6.5. System Feature Requirements

6.5.1. Responsibility of respondent: Respondents should indicate whether their proposed system for PAR, Inc., has the following features. Where features are optional, retail prices (unit cost) should be listed.

6.5.1.1. Abandoned Call Search

6.5.1.2. Abbreviated Dialing

6.5.1.3. Adjunct/Switch Applications Interface (Administered Connections)

6.5.1.4. Agent Call Handling

6.5.1.5. Alphanumeric Dialing

6.5.1.6. Voice-Activated Calling

6.5.1.7. Attendant Auto-Manual Splitting Attendant Call Waiting

6.5.1.8. Attendant Control of Trunk Group Access

6.5.1.9. Attendant Direct Extension Selection with Busy Lamp Field

6.5.1.10. Attendant Direct Trunk Group Selection

6.5.1.11. Attendant Display

6.5.1.12. Attendant Recall

6.5.1.13. Attendant Release Loop Operation

6.5.1.14. Automated Attendant

6.5.1.15. Automatic Callback

6.5.1.16. Automatic Call Distribution

6.5.1.17. Automatic Incoming Call Display (including the second call)
6.5.1.18. Automatic Wake-up
6.5.1.19. Basic Call Management
6.5.1.20. Bridged Call Appearance Multiline and Single Line Sets Busy Verification of Terminals and Trunks
6.5.1.21. Call By Call Service Selection
6.5.1.22. Call Coverage
6.5.1.23. Call Forward All Calls
6.5.1.24. Call Park
6.5.1.25. Call Pickup
6.5.1.26. Call Prompting
6.5.1.27. Call Vectoring
6.5.1.28. Call Waiting Termination
6.5.1.29. Centralized Attendant Service
6.5.1.30. Class of Restriction
6.5.1.31. Class of Service
6.5.1.32. Code Calling Access
6.5.1.33. Conference–Attendant 6 party
6.5.1.34. Coverage Incoming Call Identification
6.5.1.35. Default Dialing
6.5.1.36. Dial Access to Attendant
6.5.1.37. Direct Inward Dialing
6.5.1.38. Direct Outward Dialing
6.5.1.39. Distinctive Dialing
6.5.1.40. Do Not Disturb
6.5.1.41. Digital Carrier Service (T1, ATM, VAN, etc.)
6.5.1.42. Emergency Access to Attendant
6.5.1.43. Facility Busy Indication
6.5.1.44. Feature Transparency Between Networked Systems
6.5.1.45. Flexible Entry of Account Codes
6.5.1.46. Hold
6.5.1.47. Hunting
6.5.1.48. Inbound Call Management
6.5.1.49. Individual Attendant Access
6.5.1.50. Integrated Directory
6.5.1.51. Integrated Services Digital Network Basic Rate Interface
6.5.1.52. Integrated Service Digital Network Primary Rate Interface
6.5.1.53. Intercept Treatment
6.5.1.54. Intercom_Automatic
6.5.1.55. Intercom_Dial

6.5.1.56. Inter-PBX Attendant Calls
6.5.1.57. Interchange Carrier Access
6.5.1.58. Last Number Dialed
6.5.1.59. Loudspeaker Paging Access
6.5.1.60. Manual Message Waiting
6.5.1.61. Manual Originating Line Service
6.5.1.62. Manual Signaling
6.5.1.63. Multiple Appearance Preselection and Preference
6.5.1.64. Multiple Listed Directory Numbers
6.5.1.65. Music-on-Hold Access
6.5.1.66. Names Registration
6.5.1.67. Night Service — Hunt Group
6.5.1.68. Night Service — Night Console Service
6.5.1.69. Night Service — Night Station Service
6.5.1.70. Night Service — Trunk Answer From Any Station
6.5.1.71. Night Service — Trunk Group
6.5.1.72. Off-Premises Station
6.5.1.73. Personal Central Office Line
6.5.1.74. Personalized Ringing
6.5.1.75. Power Failure Transfer
6.5.1.76. Priority Calling
6.5.1.77. Privacy — Attendant Lockout
6.5.1.78. Privacy — Manual Exclusion
6.5.1.79. Property Management System Interface
6.5.1.80. Queue Status Indications
6.5.1.81. Recall Signaling
6.5.1.82. Recorded Announcement (with options and priority)
6.5.1.83. Recorded Telephone Dictation Access
6.5.1.84. Remote Access with Security Features
6.5.1.85. Report Scheduler and System Access
6.5.1.86. Restriction — Controlled
6.5.1.87. Restriction — Miscellaneous Terminal
6.5.1.88. Restriction — Miscellaneous Trunk
6.5.1.89. Restriction — Toll/Code
6.5.1.90. Restriction — Toll
6.5.1.91. Restriction — Telephone — Inward
6.5.1.92. Restriction — Telephone Handset — Manual Terminating Line
6.5.1.93. Restriction — Telephone Handset — Origination
6.5.1.94. Restriction — Telephone Handset — Termination
6.5.1.95. Ringback Queuing
6.5.1.96. Ringer Cutoff
6.5.1.97. Rotary Dialing
6.5.1.98. Security Violations Notification

6.5.1.139. Automatic Route Selection

6.5.1.140. AAR/ARS (Automatic Alternate Routing/Automatic Route Selection) Partitioning

6.5.1.141. Nx64 (ATM)

6.5.1.142. e remotely like outsourcing vendor)

6.5.1.143. Skills Based Routing

6.5.1.144. Stroke Counts

6.5.1.145. TCP/IP Support

6.5.1.146. Voice Response Integration

6.5.1.147. CTI API support (JTAPI — Java Telephony Application Interface, JSAPI — Telephony Server Application Interface, TAPI — Telephony Application Programming Interface) over multiple client/ server platforms

6.5.1.148. IVR

6.5.1.149. Fax on Demand

6.5.1.150. Bridged Call Appearance — Single-Line Station Call Coverage (allows multiple appearance telephone handset to have appearance of a single-line telephone handset's extension number)

6.5.1.151. Consult (allows a covering user to call called party for private consultation)

6.5.1.152. Coverage Callback (allows a covering user to automatically leave a message for the called party to call the calling party)

6.5.1.153. Coverage Incoming Call Identification (allows multi-appearance telephone handset users without a display in a coverage answer group to identify an incoming call to that group)

6.5.1.154. Go to Cover (allows users to send the call directory to coverage when making a call to another internal extension)

6.5.1.155. Send All Calls (allows users to temporarily direct all incoming calls to a coverage point)

6.5.1.156. Conference — Multi-Line Station

6.5.1.157. Conference — Single-Line Station

6.5.1.158. Distinctive Ringing

6.5.1.159. Extension Preselection and Preference (provides multi-appearance telephone users with options for placing or answering calls on selected appearances)

6.5.1.160. Hold (allows telephone users to disconnect from a call temporarily, use the telephone for other purposes)

6.5.1.161. Security Features

6.5.1.162. Toll Fraud Control

6.5.1.163. Security Audit Log including Security Violation Notification

6.5.1.164. Password Control

6.5.1.165. System Admin.'s Security

6.5.1.166. Remote Access Security

6.5.1.167. Physical Security — Console

6.5.2. Least Cost Routing

6.5.2.1. The system must provide software to route long-distance calls over the least costly available trunk group.

6.5.2.2. The proposed system must be able to change automatically the routing plan as often as required.

6.5.2.3. Each routing should able to support up to 256 routing patterns.

6.5.2.4. Up to six trunks must be programmable per routing pattern.

6.5.2.5. The routing patterns should be able to support local CO, foreign exchange, tie trunks, service, SDN trunks, and ISDN trunks.

6.5.2.6. The system administrator must be able to override an established schedule for special situation such as low holiday toll rates.

6.5.2.7. The attendant must be able to control the trunk group access from the attendant console.

6.5.2.8. International call routing must be supported.

6.5.2.9. Queuing must be provided in case all trunk groups in a routing pattern are unavailable.

6.5.2.10. Certain routes for call completion must be permitted only when a valid authorization code is entered by the station user.

6.5.3. Attendant Features

6.5.3.1. Call Waiting

6.5.3.2. Centralized Attendant Service

6.5.3.3. Control of Trunk Group Access

6.5.3.4. Direct Extension Selection with Busy Lamp Field Display

6.5.3.5. Recall

6.5.3.6. Release Loop Operation

6.5.4. Station Features

6.5.4.1. The proposed system software must provide the following station features:

6.5.4.2. Bridge Call

6.5.4.3. Call Coverage

6.5.4.4. Call Forwarding All Calls

6.5.4.5. Display Telephone

6.5.4.6. Distinctive Ringing

6.5.4.7. Hold

6.5.4.8. Last Extension Dialed

6.5.4.9. Last Number Dialed

6.5.4.10. Line/Feature Status Indication Multi-appearance Preselection and Preference Override

6.5.4.11. Priority Calling

6.5.4.12. Ring Transfer

6.5.4.13. Ringer Cutoff

6.5.4.14. Terminal Busy indications

6.5.4.15. Transfer

6.5.4.16. Trunk Verification Telephone

6.5.5. Data Support Requirements

6.5.5.1. The proposed system should support simultaneous voice and data transmission from a single port.

6.5.5.2. The proposed system must support stand-alone data applications.

6.5.5.3. The proposed system must support asynchronous data transmission at speeds up to 56Kbps.

6.5.5.4. The proposed system must transmit the data transparency.

6.5.5.5. Describe how the proposed system handles data switching.

6.5.5.6. Describe system resource requirement (processor overhead) for a data call.

6.5.5.7. Describe how data capability can be added to a telephone.

6.5.5.8. Terminal keyboard dialing must be supported.

6.5.5.9. ATD (Attention Dial the phone) command must be supported.

6.5.5.10. Autobaud (automatic speed recognition) and autoparity (automatic parity check) must be supported.

6.5.5.11. Data privacy to all data switch connections must be provided.

6.5.5.12. The establishment of permanent data switching connections between data endpoints must be provided.

6.5.5.13. DMI (Digital Multiplexed Interface) must be supported for host computer connections.

6.5.5.14. Describe what data rate interfaces are available.

6.5.5.15. Can both PVC (Permanent Virtual Circuit) and SVC (Switched Virtual Circuit) calls be made?

6.5.6. Outcalling
    6.5.6.1. The following administrator capabilities must be allowed:
    6.5.6.2. Turn outcalling on/off for the whole system.
    6.5.6.3. Allow outcalling by class of service.
    6.5.6.4. Allow outcalling on per-subscriber basis.
    6.5.6.5. Administer retry interval.
    6.5.6.6. Set a limit on the number of simultaneous outcalls.
    6.5.6.7. Control the number of digits in an outcalling number.
    6.5.6.8. Specify when outcalling is allowed.
    6.5.6.9. The following subscriber capabilities should be allowed and changeable from any touch-tone phone:
        6.5.6.9.1. Specify outcalling number.
        6.5.6.9.1. Specify when to be outcalled.
        6.5.6.9.1. Turn outcalling on/off.
        6.5.6.9.1. Specify pager number and have outcalls go to a digital pager number and have outcalls go to a voice pager.
        6.5.6.9.1. Cancel further outcalls until a brand new message arrives.
        6.5.6.9.1. Restrict outcalls only if an urgent message is received.

6.5.7. Automated Attendant
    6.5.7.1. The following capabilities should available for the automated attendant:
        6.5.7.1.1. Multiple menu choices (state maximum).
        6.5.7.1.2. Extension can be entered in addition to menu choices.
        6.5.7.1.3. Callers can specify who they want to contact by name rather than extension.
        6.5.7.1.4. Multiple levels are allowed in an automated attendant (i.e., make a choice and be presented with another set of choices).
        6.5.7.1.5. A default can be established if callers do not enter any choice (i.e., time out to an extension).
        6.5.7.1.6. The administrator can record the attendant menu from any touch-tone phone.

6.5.7.1.7. The menu can be repeated to callers either by pressing a button or timing out.

6.5.7.1.8. A caller can specify an invalid choice or extension, and can be notified and given another opportunity to make a choice.

6.5.7.1.9. Several callers can be in the same automated attendant simultaneously.

6.5.7.1.10. There can be at least 100 automated attendants defined in the system.

6.5.7.1.11. A second attendant menu is available for after-hours.

6.5.7.1.12. A different attendant menu can be played to external callers than the one played to internal callers.

6.5.7.1.13. A message can be left in the attendant mailbox.

6.6. Physical Requirements

6.6.1. Footprint

6.6.1.1. Available space is approximately 40' × 30' of raised flooring.

6.6.2. Weight Requirements

6.6.2.1. In some cases, heavy equipment requires additional structural support on the 51st floor of the PAR, Inc. building. Vendor should be prepared to include in the price any cost of facilities upgrade to support additional weight (especially in view of the additional weight expected during the transition, when both the old and new PBXs will be on the 51st floor).

6.6.3. Shipping

6.6.3.1. Shipping will be FOB destination. Vendor assumes risk of loss until equipment is physically inside the PAR, Inc. building.

6.6.4. System Power

6.6.4.1. Describe specification of power supply.

6.6.4.2. The proposed system must be properly grounded to protect against the effects of ground loops, pickup noise, and excessive ground current.

6.6.4.3. Standby Power System

6.6.4.4. Describe requirements for standby power needs.

6.6.4.5. Describe the battery backup options available with proposed system. Include a list of a backup battery options.

6.6.4.6. The standby power system must support AC and DC load requirements for equipment installed in PAR, Inc. /Chattanooga building.

6.6.4.7. Describe what happens if the AC failure lasts long enough to fully discharge the batteries.

6.6.4.8. Describe alarm indication and monitoring capability available for the standby power system.

6.6.5. Equipment Room

6.6.5.1. Provide a proposed floor plan showing the equipment room layout, including the floor space, HVAC, and electrical requirements.

6.6.5.2. Identify the Main Distribution Frame locations.

6.6.5.3. Identify the placement, dimensions, and configuration of all cabinets.

6.6.5.4. Consideration of space limitation for installation of the proposed main switch (must be able to fit into the existing floor space).

6.6.5.5. Specify the ground requirements for the proposed system and other ancillary equipment (i.e., voice mail system).

6.6.5.6. Specify the <u>recommended</u> room temperature and humidity ranges for the proposed system.

6.6.5.7. Specify temperature and humidity ranges <u>acceptable</u> to the proposed system.

6.6.5.8. Specify the heat dissipation BTUs per hour of the proposed system.

6.7. Multimedia Integration Requirements and Voice Mail

6.7.1. Personal Features and Desktop Conversion

6.7.1.1. The respondent should describe the costs and effort required to convert PAR, Inc.'s desktops (PAR, Inc. Knoxville and Chatanooga buildings) to use unified messaging. Since the number of conversions is unknown, a rate per desktop should be stated.

6.7.1.2. Rates to convert personal features (e.g., personal distribution lists) should also be listed. If this is bundled with the entire installation, individual rates are not required for this item.

6.7.2. LAN (Local Area Networking)

6.7.2.1. Describe how the proposed system supports and interfaces with LANs.

6.7.2.2. Does the proposed system support Ethernet, fast Ethernet, and Token Ring LANs?

6.7.2.3. Does the proposed system support ATM?

6.7.2.4. What network protocols are supported?

6.7.2.5. Does the proposed system support LAN inter-
working via bridging or routing?

6.7.2.6. Is the internetworking fully integrated, utilizing
shared bandwidth of the digital carrier (T1,
Frame Relay, ATM, SMDS, etc.)?

6.7.2.7. What form of LAN network management is implemented?

6.7.2.7.1. Is the network management system
compliant with an industry standard
(SNMP, CMIP, etc.)?

6.7.2.7.2. Describe any LAN/WAN security features.

6.7.2.7.3. E-mail integration capability (must
be current not future) with Lotus
Notes version 5.0 and above, and
compatibility with browser.

6.7.3. Message Creation (Voice Mail)

6.7.3.1.1. Re-record from any place within the
message.

6.7.3.1.2. Pause during message creation.

6.7.3.1.3. Go forward and backward within the
message in small steps.

6.7.3.1.4. Review before sending.

6.7.3.1.5. Mark messages as urgent.

6.7.3.1.6. Notification that a message was successfully delivered.

6.7.3.1.7. Notification that a message was accessed by the recipient.

6.7.3.1.8. Notification of a successful network
delivery.

6.7.3.1.9. Notification that a subscriber on a
remote system (network) accessed
the message.

6.7.4. Other Voice Mail Features

6.7.4.1.1. The proposed system must protect
all messages, data, and software during outages.

6.7.4.1.2. The equipment must be UL ap-
proved.

6.7.4.2. Technical Specifications

6.7.4.2.1. Does the proposed system use automatic gain control to adjust volume
levels?

6.7.4.2.2. The proposed system must restart automatically once power is restored after a power outage.

6.7.4.2.3. Diagnostics must run 24 hours per day without system disruption.

6.7.4.2.4. Does the voice-encoding algorithm save disk space by not recording silences onto the disk?

6.7.4.2.5. Can all data including personal greeting, recorded names, system software, and all messages automatically be backed up at least once a day?

6.7.4.2.6. Can this same critical data be duplicated in realtime so an "instantaneous" copy is available?

6.7.4.2.7. The system must allow for mailbox numbers to be changed without losing messages, mailing lists, or subscriber's names.

6.7.4.2.8. Does the proposed system provide a history log that records system problems?

6.7.4.2.9. Does voice mail smoothly interface with the Hoplitelog recording system? Does it interface with any other third-party recording systems (comprehensive list not required — just a few representative recording systems or your own recording system)?

6.7.4.2.10. Does the proposed system alert the administrator when message space is low?

6.7.4.2.11. The proposed system must alert subscribers when their mailbox space gets low.

6.7.4.2.12. HELP prompts must be provided to the users.

6.7.4.2.13. Does the proposed system provide for multiple security levels for system administration?

6.7.4.2.14. Can System Administration remotely access features?

6.7.4.2.15. Does the proposed system have an inherent redundancy?

6.7.4.2.16. Can subscribers and/or administrators record their own name?

6.7.4.3. Feature Specifications

6.7.4.3.1. Call Answer

6.7.4.3.1.1. Message Creation Calls must be automatically answered by the called party's greeting without reentering the mailbox number.

6.7.4.3.1.2. A caller must be automatically routed out of the voice mail system to an attended telephone by pressing a single digit.

6.7.4.3.1.3. Can each mailbox have a unique destination in the above scenario?

6.7.4.3.1.4. All subscribers must be able to "escape/transfer out" of voice mail.

6.7.4.3.1.5. Callers must be able to transfer out to any extension.

6.7.4.3.1.6. Callers must be able to transfer out by specifying a subscriber's name.

6.7.4.3.1.7. Callers must be able to first leave a message and then transfer out to another extension or the attendant.

6.7.4.3.1.8. Callers must be able to re-record their message.

6.7.4.3.1.9. Callers must be able to skip the greeting and immediately record message.

6.7.4.3.1.10. The caller must be able to mark the message as private.

6.7.4.3.1.11. Can callers who are also subscribers, transfer into their own mailbox after leaving a message or in place of leaving a message?

6.7.4.3.1.12. If a caller is within the same switch, will the recipient get the caller's extension and name when retrieving the message?

6.7.4.3.1.13. What is a caller's maximum message time?

6.7.4.3.1.14. The system administrator must be able to vary the message length on a per-subscriber basis.

6.7.4.3.1.15. The proposed system must provide subscribers with the ability to greet their callers with a personal message.

6.7.4.3.1.16. If the called party's mailbox is full, will the system tell callers what their other available options are?

6.7.4.3.1.17. Can subscribers choose to use a generic system greeting rather than a personalized greeting?

6.7.5. Message Addressing/Scheduling

6.7.5.1. The system must provide the following capabilities to subscribers during addressing or scheduling messages for delivery:

6.7.5.1.1. Address by extension.

6.7.5.1.2. Address by name.

6.7.5.1.3. Address some by extension and others by name.

6.7.5.1.4. The system voice plays back the name.

6.7.5.1.5. If the addressee is on a remote system (networked), the name must be voiced.

6.7.5.1.6. Address to a predetermined mailing/distribution list.

6.7.5.1.7. Mailing lists can be created and maintained by subscribers.

6.7.5.1.8. Mailing lists can be owned by one person but used by others.

6.7.5.1.9. If one person appears on two or more mailing lists that are sent the same message, does the subscriber only get the message once?

6.7.5.1.10. The system allows future deliveries of up to one year in the future.

6.7.5.1.11. A future delivery can be cancelled any time before delivery.

6.7.5.1.12. The message can be marked private during the addressing or scheduling sequence.

6.7.5.1.13. Addresses can be removed prior to sending the message, even if they were part of a predefined mailing list.

6.7.5.1.14. Subscribers can set up aliases for shortcut way to address by name.

6.7.5.1.15. A message can be broadcast to everyone without using a mailing list.

6.7.5.1.16. The System Administrator can maintain system distribution list, which can be accessed by any subscriber.

6.7.6. Message Retrieval

6.7.6.1. The system must give the number of new message at log-in time.

6.7.6.2. Are saved messages stored in a different category from new message?

6.7.6.3. Can the categories be presented in any order?

6.7.6.4. Can message be played back first-in, first-out, and can they also be played back last-in, first-out?

6.7.6.5. Can a voice mail subscriber elect to skip to next message category?

6.7.6.6. Can a voice mail subscriber elect to skip to old message?

6.7.6.7. Can a voice mail subscriber cancel review of messages?

6.7.6.8. Can a voice mail subscriber replay messages?

6.7.6.9. Can a voice mail subscriber delete message at any time?

6.7.6.10. Can a voice mail subscriber review message header information?

6.7.6.11. Can a voice mail subscriber pause during review?

6.7.6.12. Are speed controls available?

6.7.6.13. Are volume controls available?

6.7.6.14. Can the system play one message after another without intervention?

6.7.6.15. Are messages presented before all other new messages?

6.7.6.16. Are broadcast messages resorted before urgent and all other new messages?

6.7.6.17. Does the system tell subscribers they have new broadcast messages at log-in time?

6.7.6.18. Does the system tell subscribers they have new urgent message at log-in time?

6.7.6.19. Is the following information available for each message?

    6.7.6.19.1. Sender's name

    6.7.6.19.2. Sender's extension

    6.7.6.19.3. Date and time message was delivered

    6.7.6.19.4. Type of message (private, internal, external)

    6.7.6.19.5. Private status (only if this is a private message)

    6.7.6.19.6. Urgent status (only if this is an urgent message)

6.7.6.20. Can a voice mail subscriber reply to sender, without re-addressing?

6.7.6.21. Can a voice mail subscriber add a comment to the beginning or the end of the message and forward?

6.7.6.22. Can a voice mail subscriber record a new message and send without losing place in your message queue?

6.7.6.23. Can a voice mail subscriber have multiple responses to one message?

6.7.6.24. Can a message be sent from one subscriber to multiple extensions, names, or distribution lists and still retain all added comments?

6.7.7. Networking

6.7.7.1. When addressing a message to remote subscribers, is the name displayed back for verification?

6.7.7.2. Can subscribers who send messages to remote subscribers find out whether those messages have been delivered and, later, that they have been accessed?

6.7.7.3.   Can predefined mailing lists contain both local and remote subscribers?

6.7.7.4.   If a subscriber's name changes or extension number changes and is updated on the home system, does this get updated automatically on all remote systems?

6.7.8.   Voice, Fax, E-mail on Desktop

6.7.8.1.   Can voice, fax, and e-mail be integrated onto the desktop?

6.7.8.2.   Can e-mail by read over the phone by the phone mail system to users (i.e., by a "robot" voice)?

6.7.8.3.   Does the system require hardware/cabling to connect the PC to the telephone or is the link at the sever?

6.7.8.4.   Can the system OCR a fax and read it back to the user over the phone?

6.7.8.5.   Can voice messages be replayed over the phone and over PC speakers?

6.7.8.6.   Do you have the facility to use voice recognition to connect a call?

6.7.8.7.   Describe the formats (e.g., "wav" files) for user accessible voice, fax, and e-mail files. Can they be stored locally?

6.8.   System Management, Administration, and Reporting Requirements

6.8.1.   Product Requirements

6.8.1.1.   Specify the operating system required as well as software and database language.

6.8.1.2.   Specify the hardware and software required of the proposed management system.

6.8.1.3.   The proposed system must be capable of supporting PAR, Inc.'s functional requirements with the capacity of growth.

6.8.2.   Network Level Traffic Measurement Reports

6.8.2.1.   Can the system collect traffic and call detail data from more than one switch?

6.8.2.2.   Specify the capacity of your proposed traffic measurement system.

6.8.2.3.   Does the traffic management application support the customized reporting?

6.8.2.4.   Explain how the system provides for traffic management of the system's trunk and trunk group traffic.

6.8.2.5.   Explain system administration of the network traffic system.

6.8.2.6.   Explain what reports are available to perform trend analysis using historical data.

6.8.2.7.   Explain what reports are available through ACD (Automatic Call Distribution System).

6.8.3.   Traffic Management and Reporting

6.8.3.1.   Can the system collect traffic data to help monitor the system utilization?

6.8.3.2.   Explain how the system provides for traffic management of the system such as common equipment usage, individual attendant traffic, attendant traffic, trunk traffic, trunk group traffic, feature usage, and security violations.

6.8.3.3.   Explain the reporting periods for which reports are available.

6.8.3.4.   Explain how the available traffic reports can be used to evaluate traffic trend and ensure a sound traffic flow.

6.8.4.   Reporting Capabilities

6.8.4.1.   The proposed system must have the ability to run reports on demand or scheduled intervals for printed reports.

6.8.4.2.   Include copies of reports available through the system administration terminal, including system measurement, system status, and historical reports.

6.8.5.   Administration of Components

6.8.5.1.   Provide a list of the switch components and features that can be administered with the proposed system.

6.8.5.2.   The administrative tool must support multiple users and be able to administer the switch remotely.

6.8.6.   Systems Management

6.8.6.1.   The proposed system should be designed to provide realtime access to administrative procedures (both on-site and remote access).

6.8.6.2.   The proposed system should provided the ability to view online data associated with the switch.

6.8.6.3.   The database for the management system must be automatically updated when MAC (churn) is performed.

6.8.6.4.   The proposed management system should provide access to maintenance procedures for testing and troubleshooting stations and trunks.

6.8.6.5. Security of the system must be maintained through passwords and logon-id.

6.8.6.6. The management system must provide realtime online update with full-screen editing.

6.8.6.7. Extensive activity log files for tracking and reconstructing a sequence of events must be inherent in the system management.

6.8.7. System Management (Voice Mail)

6.8.7.1. The following capabilities should be available:

6.8.7.1.1. Mailboxes can be added, deleted, or changed without service interruption.

6.8.7.1.2. Retention times can be established so that messages are automatically deleted after a prescribed number of days.

6.8.7.1.3. The administrative interface is forms-driven.

6.8.7.1.4. A subscriber's extension can be changed without deleting their messages or changing anything else in their mailbox.

6.8.7.2. The following parameters should be able to be modified on a class of service and per-subscriber basis:

6.8.7.2.1. The mailbox size (number of messages) is:

6.8.7.2.2. The length of message a caller can record (call answer) is:

6.8.7.2.3. The length of message a subscriber can record (voice mail) is:

6.8.7.2.4. The retention time on a new message is:

6.8.7.2.5. The retention time of old messages is:

6.8.7.2.6. Outcalling is allowed.

6.8.7.2.7. The privilege exists to mark messages as urgent.

6.8.7.2.8. Ability exists to broadcast messages to everyone without using a mailing list.

6.8.7.3. The following statistics should be provided by the system:

6.8.7.3.1. Total number of calls for each port, and total seconds busy for each port.

6.8.7.3.2. Total messages created.

6.8.7.3.3. Disk utilization.

6.8.7.3.4. Number of subscribers and automated attendants.

6.8.7.3.5. Average call answer session (in minutes and/or seconds)

6.8.7.3.6. Subscribers directory by extension and name.

6.8.7.3.7. List of all automated attendants.

6.8.7.3.8. Average ports in use by hour.

6.8.7.3.9. The following subscriber statistics should be provided:

6.8.7.3.10. Message created.

6.8.7.3.11. Successful log-ins.

6.8.7.3.12. Failed log-in attempts.

6.8.7.3.13. Messages received (call answer).

6.8.7.3.14. Messages received (voice mail).

6.8.7.3.15. Maximum space used per mailbox.

6.8.7.3.16. The following network statistics should be provided:

6.8.7.3.17. Number of messages sent between two systems.

6.8.7.3.18. Number of network connects between two systems.

6.8.7.3.19. Number of failures to each network system.

6.8.7.4. Security

6.8.7.4.1. Subscribers can create or change their personal password at any time, from any touch-tone telephone.

6.8.7.4.2. The system prevents the administrator from obtaining personal passwords.

6.8.7.4.3. The administrator can give a subscriber a new password without obtaining the old password.

6.8.7.4.4. New subscribers are required to change the default password upon initial logon into the system.

6.8.7.4.5. Variable length password supported.

6.8.7.4.6. A minimum length can be set for passwords.

6.8.7.4.7. The system disconnects after three unsuccessful log-in attempts (mailbox should lock).

6.8.7.4.8. The system can require subscribers to change their passwords at pre-defined time intervals.

6.8.7.4.9. The system can track invalid attempts to log in and notify the administrator.

6.8.7.4.10. The system requires a password to log onto the control console and parameters.

6.8.7.4.11. General, systemwide security reports showing who has not changed their password, last time a mailbox was used, high frequency of lock-outs, etc., should be available.

6.8.7.4.12. The respondent should list any solutions for encrypted voice capability through the switch. PAR, Inc., currently uses the RTY44 3600, which requires a similar station "box" on both sides of the conversation. This is not an ideal solution because of the proliferation of these devices. Ideally, encryption (specify the "bit" level) should be available through the PBX to another PBX similarly equipped. Thus, security could be handled at a system rather than an individual handset level.

6.9. Training Requirements

6.9.1. User Training Plan

6.9.1.1. Telephone use training

6.9.1.2. Voice mail training

6.9.2. Training
Technical training support for PAR, Inc. voice communications agents and third-party contractors. This includes detailed technical system level training to enable PAR, Inc., if it so chosen, to "self-maintain" its PBX and voice mail systems.

6.9.3. Basic Training

6.9.3.1. Explain the training plan to support the products recommended by respondents.

6.9.3.2. Station user and console attendant training

6.9.3.3. Describe your plan for station user and console attendant training.

6.9.4.   System Administrator Training
Describe your plan for system administrator training.

6.9.5.   Technician Training
Describe your plan to train PAR, Inc.'s communications technicians and third-party contractor.

6.9.6.   Instructor Training
6.9.6.1.   Describe your plan to train PAR, Inc.'s telecommunication system trainer.
6.9.6.2.   Describe the "train the trainer" course materials.
6.9.6.3.   Describe training location.

6.9.7.   Training Details
Provide details regarding training cost, duration, any other relevant details about the system administration training.

6.9.8.   Ongoing Training Support
Describe your plan to provide ongoing training support.

6.10. Maintenance Requirements

6.10.1.   Product Revisions and Updates
Describe customer notification procedure to inform update, product enhancements, revisions, etc.

6.10.2.   Training Materials
Describe the training materials you will furnish PAR, Inc. The following training materials are desirable:
6.10.2.1.   CBT (Computer-Based Training)
6.10.2.2.   Laser disk
6.10.2.3.   CD-ROM
6.10.2.4.   Videotape
6.10.2.5.   Multimedia presentation

6.10.3.   Maintenance
Provide a detailed overview of all applicable warranties.

6.10.4.   Service
6.10.4.1.   Describe service provided during warranty.
6.10.4.2.   Describe your service organization.

6.10.5.   Maintenance Staff
6.10.5.1.   Identify the makeup of the maintenance staff that will be assigned to the proposed system, citing their training experience on the proposed system.
6.10.5.2.   Post Warranty Maintenance Options. Please provide detailed information for these options.
6.10.5.3.   Describe your maintenance options available after the warranty period.

6.10.6.   Response Time
What are your company's response times to major and minor system failures? List, among other items, your firm's maximum time to respond to an outage.

6.10.7. Inventory
What inventories do you maintain in order to assure prompt response to repair requests?

6.10.8. Preventive Maintenance
Does the maintenance contract include provisions for preventive maintenance? Does the system have to be taken out of service?

6.10.9. Disaster Recovery Plan
Vendor must provide a plan for repair or replacement of the system in the event of a catastrophe.

6.10.10. Dedicated Technician
Describe an optional plan of a dedicated technician reporting to and working full-time at PAR, Inc.

6.10.11. Remote Maintenance
Discuss your remote monitoring, diagnostic and repair capabilities, focusing on your ability to quickly and accurately identify and resolve reported troubles.

6.10.12. Trouble Reporting
Explain your established trouble reporting procedures including a trouble report telephone number to answered 24 hours a day, seven days a week.

6.10.13. System Maintenance Billing
Explain the standard billing period for system maintenance (monthly, quarterly, annually, etc.).

6.10.14. Upgrades and Additions Procedures
Describe your procedures for software updates and upgrades.

6.10.15. Escalation Procedures
6.10.15.1. Describe your escalation procedures.
6.10.15.2. Discuss the capability of the proposed system to automatically call for help when alarm conditions occur.

6.10.16. Customer Participation
Discuss what options are available should PAR, Inc., participate in part of the maintenance process.

6.10.17. Remedial Maintenance
Discuss your remedial maintenance procedures and how often they are performed.

7. RESPONDENT OPTIONAL SECTION

7.1. Areas Not Covered by the RFP
If there are facts/considerations not addressed by this RFP that respondent feels should be included, please list below. This section is strictly optional.

8. PAR, INC.'S RESPONSIBILITIES

8.1. Outline of Responsibilities

8.1.1. The respondent should outline in detail all implementation and ongoing operational responsibilities that PAR, Inc. is expected to assume. Issues to be addressed include, but are not limited to:

    8.1.1.1. Project management

    8.1.1.2. Training

    8.1.1.3. Transition and conversion of data from the existing systems

    8.1.1.4. Change control

    8.1.1.5. Interface with LEC and IXC carriers

9. MISCELLANEOUS

9.1. TELCO call detail

9.1.1. Interface with existing TELCO call detail system.

9.1.2. Vendor should interface with TELCO call detail system.

10. FINANCIAL OPTIONS

10.1. Lease/buy options

Respondent should provide PAR, Inc., with financial options that will minimize PAR, Inc.'s cost over the expected life of the equipment.

10.2. Payment Schedule

Respondents should provide an anticipated payment schedule (separate for each option presented).

10.2. Future Purchases

PAR, Inc., expects the purchase of new telephony systems for its Corporate Headquarters to establish a direction (but not an absolute requirement) for all business units. Therefore, the same levels of discount on future purchases would be expected. Please state your position on this issue.

Footnotes

1. A fictional company.
2. Scheduled downtime must be rare (example: once per year). Monthly scheduled downtime, for example, is unacceptable.
3. Should be a field that can be modified by an PAR, Inc., switch technician on a system level.

Appendix B
Sample System Parameters

THE FOLLOWING TABLE PROVIDES A LIST OF SYSTEM PARAMETERS, many of which are drawn from the Lucent Technologies Definity G3R communications server documentation. This list is illustrative and does not include all parameters due to the length and vendor-specific nature of the tables (there are typically hundreds of parameters). However, all large-scale communications server manufacturers have similar decision points that must be reviewed and understood by the customer. The review of these parameters requires considerable effort from both the telephony group and the implementation committee.

Parameter	Explanation and Comments
Trunk to trunk transfer	Allows a call to come in on one trunk and go out another. For example, an executive is in London, calls her administrative assistant in Nashville, and asks to be transferred to the Houston office. This scenario works only if trunk to trunk transfer is turned on. Disallowing the feature improves security against toll fraud but is generally inappropriate for large organizations (turning it off would interfere with normal business operations). Well-implemented security systems and carrier monitoring can compensate for this exposure.
Coverage subsequent redirection	A call comes in to a telephone and no one answers. The call then goes to coverage, the predefined target of an unanswered call. Coverage is often either voice mail or someone else in the office who is taking calls for others. Subsequent redirection occurs when the first level of coverage is not available. This parameter establishes how many seconds the third, fourth, etc. party has to answer the call before it is transferred to the next coverage.
Coverage/caller response interval	Defines how long the person called has to answer his or her telephone. After this period (in seconds), the call goes to predefined coverage, such as voice mail or another telephone instrument.
Automatic callback/no answer timeout interval	A user dials an extension and receives a busy signal. He then enters a callback code and hangs up. When the called extension hangs up, the communications server rings the user that made the call. That user hears a distinctive ring and knows that the person he wanted to reach is now available. This parameter defines how long that distinctive ring will last. If the user that originally called picks up the telephone, the second party will be called. In Siemens systems, this feature is called "camp on."

COMPUTER TELEPHONY INTEGRATION

Call park timeout interval	In a hospital, doctors may receive a loudspeaker page "Call for Dr. Smith." The doctor goes to the nearest telephone, enters a code, and retrieves a call specifically for her. Prior to the page, someone has received a call for Dr. Smith and has "parked" it in the system until the call is retrieved. While the call is parked, ports and trunk capacity are tied up. This parameter defines how long the call can be parked before disconnect occurs. Values of 10 to 30 minutes are typical.
Abbreviated dial programming	Allows the use of stored number lists for quick dialing by users. Most installations will install this feature.
Terminal translation initialization (TTI)	Users or technicians can go to a telephone, enter a TTI code, and assign an extension to that specific telephone (virtual stations). All characteristics of the assigned telephone are carried to the new instrument (such as voice mail and user parameters). There are benefits and risks in implementing virtual stations. At the offices of one Big Six accounting firm in Houston, the professional staff are assigned a rollabout cart for all their files and find any available telephone for use while they are in the office. A code is entered in the Nortel telephone and the telephone assumes their extension. For staff that travels frequently, the virtual office reduces space and resource requirements. Given that an average office space in a downtown building can easily cost from $3000 to $10,000 per year, there is a clear economic incentive to share space where possible.
	On the other hand, TTI can become a mess if users forget to log-off or if they change telephones all over a building without informing the telephony group. Directories that relate employees to both extensions and room numbers must be maintained. Departmental information (for internal billing) is tied to extension. One alternative that provides the flexibility of TTI use, but also avoids the potential accounting disruption, is to allow only telecommunications personnel to use the feature. The selection committee and telecommunications group should consider this a strategic decision for the enterprise.
TTI security code	The ability to use TTI is further regulated by a password. It should be changed routinely.
Prohibit bridging onto calls with data privacy?	Modems, faxes, and other analog devices should not be disturbed in their communications functions by any other features such as system-generated tones or intrusion attempts.
Call forward override	Employee A has a "must do today" project and has his extension forwarded to employee B. However, the CEO calls employee A for lunch. With call forward override, employee B can forward the call back to employee A.
External coverage treatment for transferred incoming calls	Employee A has her calls forwarded to employee B. When an outside party calls employee A, employee B hears a different ring than when an insider calls.
Coverage of calls redirected off-net enabled	Allows employees who are working from home or other offsite locations to have their calls redirected to their current location. Unless this parameter is set to "yes," telecommuting will not work. Class of service/class of restriction is also used to control who can telecommute.
Call pickup alerting	Provides a flashing status light on a user's telephone to indicate that an extension within the user's call pickup group is ringing.

250

Temporary bridged appearance on call pickup	If a user "picks" an incoming call in her pickup group, the system can either deny access to that call by other members of the pick group, or allow it. It may be irritating, for example, to have the caller hear two individuals responding to his call.
Directed call pickup	Allows a user to pick up (answer) a call from any telephone on the switch. Obviously a feature that needs careful review.
Control outward/toll restriction intercept treatment	For selected extensions (e.g., a lobby telephone), a message such as "Outside calls cannot be made from this telephone" can be played if the telephone has toll restrictions. It is more pleasant to the caller than a warbling error tone. Note that in any major installation, there are dozens, perhaps hundreds of announcements that may have to be recorded.
Authorization codes enabled	The organization may want to require all or selected users to enter authorization codes for long-distance calls (or perhaps only international calls). The authorization codes are included in call detail records that are transmitted via the CDR port. Charges can later be sorted by authorization code and distributed to the appropriate department.
Malicious call trace warning tone	The telecom organization may be asked to identify the origin of harassing calls made to a specific extension. If call trace is set up for a specific extension, a warning tone can be generated for the user only, or for both user and caller.
Coverage	Coverage is the action taken when an extension is not answered. In many cases, coverage is simply delivering the call to voice mail after a specified number of rings. The enterprise should decide whether every extension will have coverage or some will not. If some stations have no coverage, then an interval should be established, after which a intercept tone is given.
Send ISDN trunk group name on tandem calls	While this is not a major parameter, it is illustrative of the many technical specifications that influence ease of maintenance for the switch. By sending the trunk group name, technicians diagnosing a problem can more readily identify and diagnose a bad trunk.
Attendant (telephone operator) parameters	There are a multitude of attendant specifications. Use of these will depend on the size of the organization and the role of the agents. Generally, for large organizations, the attendant's role should be limited to routine business functions rather than handling exceptions such as responding to an outside caller who has dialed a non-working DID. Care needs to be exercised so that attendants do not inadvertently get overwhelmed with exception calls.
Outpulse without tone	The switch may communicate with some devices, such as Cisco's Stratacom ATM device, that should not receive a tone before transmission.
Date format	Format for display of the date on telephones with that capability.
Misoperation alerting	What should the user hear when a wrong button is pushed on the telephone?
Service observing warning tone	Indicates when a supervisor or other individual is listening to a conversation. When making this decision, the implementation committee should obtain legal counsel and also consider carefully the culture of the organization. Silent monitoring can be considered an unwarranted invasion of privacy unless organizational policy is clearly stated and documented.

Security violation login enabled	The organization needs to define what security violations will be reported and where the notifications will be sent. They may be sent to a designated individual's pager, to the Help Desk, to a log, or to a security console. The response should be appropriate to the organization's size and the severity of the violation (e.g., users will constantly forget their voice mail passwords — having a telecom employee paged when an employee locks himself out may not be practical).
Call detail recording	Decisions need to be made regarding the format of the call detail records, and whether to record outgoing, incoming, or station-to-station calls. In some systems, station-to-station calls can be recorded for only selected extensions. Other questions include: should be "#" sign be removed from the number; when a call is transferred, what extension should be shown in the call detail record; and when a call is made to a hunt group, what extension is recorded — the hunt group number or the extension that actually takes the call.
Suppress CDR for ineffective call attempts	Callers may dial only a few digits and then hang up, or may dial the entire number but hang up before the call is completed. Generally, the records that could be generated from this activity are suppressed.
Extension length	The dial plan must be specified. The switch looks at each digit as it is dialed and must have explicit instructions to process the call. For example, it may look at the first digit, find "8" and know to send the call to tie lines.
Type device	Each physical telephone model must be entered into the switch. In some cases, a close match can be entered for analog telephones. However, digital telephones remain proprietary for most vendors so that, for example, a NEC digital telephone will not work with a Siemens PBX.
Display language	The switch must be told what language to use in displays.
Class of service and class of restriction	A detailed table must be constructed that allows an appropriate segregation of capabilities. This breakdown serves to both enhance security and promote efficient operations. For example, the CEO may have "barge-in" capabilities and the ability to do a trunk-to-trunk transfer from her home. The lobby telephone may only be able to call internally. The table must be constructed carefully so that the organization does not run out of classes to assign to special users.
	Some representative features controlled by class of service include: automatic callback, call forwarding all calls, data privacy, priority calling, restrict call forwarding off-net, call forward busy/do not answer, personal station access, off-hook alert, and console permission.
Ringing patterns	Distinctive ringing is usually a station feature rather than a system-level feature. Varying ringing patterns are most valuable in a high-density office where employees could mistake a co-worker's ring for their own.
Redirect notification	Sends a "half ring" to an employee's station when the station is forwarded to another location or voice mail.

Answer supervision	Answer supervision is important for accurate billing. When the central office sends a call to the switch, a return signal is sent by the switch when the call is answered (picked up). Without answer supervision, the CO must assume that the call is answered after an arbitrary amount of time and hence billing is inaccurate. Answer supervision should be turned on.
Automatic route selection (ARS)	Routes calls over the most economical path. For example, the switch looks at the number dialed and, if applicable, sends it along a tie line rather than the public network.
Audible message waiting	For instruments lacking a message waiting light, a stutter tone at the beginning of dial tone indicates a message has been left in voice mail. The feature is also valuable for visually impaired individuals.
Priority calling	Allows the use of distinctive three-burst ringing to call an extension.
Forced entry of account codes	The switch can require all long-distance calls to have a valid account code entered before the call can go through.
Personal Station Access	For situations where employees will be routinely sharing a telephone (e.g., two employees work part-time at the same desk), a personal station access code can be entered that allows each one to be uniquely identified with separate call features, etc. Each may have their own voice mail, coverage paths, and class of service. It is a high-maintenance feature from the perspective of the telecommunications staff, and therefore should be used only where it is really needed.
Tenant partitioning	In environments where the telephony system is shared by unrelated tenants (e.g., an office building not occupied by a single entity), tenant partitioning can be used to segregate billing and other services, such as music on hold. All parties are served out of the same switch.

Appendix C
Sample Service-Level Agreement: Pallas Athene Reproductions, Inc.

SERVICE-LEVEL AGREEMENT FOR VOICE COMMUNICATIONS SERVICES between the Voice Communications Group (VCG) and the Knoxville Call Center, Department #41.

HISTORY OF CHANGES

Revisions and release dates of this document are documented below:

Document Version	Date	Author	Comments
1.0	11/2/1999	Jane Sappho	Initial implementation document.
1.1	12/15/1999	Bill Scopas	Changed after hours response time from 2 hours to 1 hr, 30 minutes

CONTENTS

SERVICE-LEVEL AGREEMENT (SLA) GOALS

1. **Communication**
   Ensure that the customer and service provider communicate effectively and set up appropriate expectations.
2. **Resolve disputes**
   Provide a method for avoiding disputes by providing a shared understanding of needs and priorities.
3. **Up-to-date Document**
   Review the agreement periodically to ensure that it continues to be relevant and meet requirements. Makes notes of meetings and agreements as appropriate.
4. **Scorecard**
   Establish a written scorecard for evaluating service success. Both provider and customer should agree on the basis of the evaluation.

Service Objectives

1. Provide telephone connectivity for Call Center, Department #41 users.
2. Provide 99.997 percent availability for PBX, CTI, IVR, and voice mail communications.

ORGANIZATIONAL INFORMATION

Customers

1. Call Center, Department #41
   A. Roxanne Alexander

Service Providers/Enablers

1. Common dispatch center
   A. Fred Plato
2. Voice Communications Group (VCG)
   A. Jane Sappho
   B. James Smyth
   C. Bill Scopas
3. PARS Information Systems (PARS-IS)
   A. Patricia Pericles

Responsibilities by Business Group

Customer

1. Provide a coordinator for a single point of contact (SPOC).
2. Report any service problems to VCG Help Desk. Assign resources as needed to collaborate with VCG in problem resolution.
3. Maintain consistent priority levels applicable to telephony service problems with VCG.

4. Provide VCG with notification of needs that will materially affect the voice communications system, including CTI, IVR, and other adjunct processors and applications. Notification should be given in order to provide sufficient time to analyze impact to telephony systems.
5. Maintain general communications to ensure that the service provider is not "blindsided" by unanticipated and radical changes in volume requirements or new services.

Provider
1. Voice Communications Group
   A. Will be responsible for the proper installation, maintenance, and performance of the voice communications system. This includes all handsets, PBX components, voice mail equipment, and any adjunct processors housing voice applications such as IVR or CTI. Track and manage all reported problems.
   B. Provide support at resource levels comparable to staffing previously dedicated to Call Center, Department #41, and commensurate with the staffing budget transferred to VCG.
      (1) Remote administration of sites outside Knoxville from 7 a.m. to 6 p.m. Monday–Friday. Dispatch of service representative to remote sites within four hours of request for those sites with no VCG staffing.
      (2) On-site coverage for the PARS Building and Building 4A from 6 a.m. until 7 p.m. Reported problems are responded to within 15 minutes during coverage hours. Response is 1.5 hours during non-coverage hours.
      (3) Manage trunks, T1s, ADSL lines, and other carrier facilities necessary to carry voice traffic.
      (4) Manage Centrex, if desired.
      (5) Report call detail via remote poll using ABC Call Accounting system; call buffer box must be purchased by customer.
      (6) Provide training materials for end users (how to use features, etc.). Include both telephone and voice mail features.
      (7) Make telephone modifications requested by users (pick group, forwarding, name display, hunt group, line appearances, etc.)
      (8) Provide for moves, adds, changes. Additional charges apply if vendor must send technician for site visit.
      (9) Man the Help Desk between 7 a.m. and 6 p.m.
      (10) Provide consultation, e.g., special projects such as fax-on-demand, ACD, CTI programming, and IVR. Additional charge at hourly rate.
      (11) Add sites that have VPIM voice mail compatibility, or native voice mail interface with Lucent, will be added to the corporate voice mail directory, allowing transfer of messages from

PARS Building and Knoxville, Building 4A to those sites via Intuity Interchange platform.
C. Conditions of Service Availability
(1) Dependency on carrier networks for providing local and
long-distance connectivity. In those cases where dedicated
lines and routing equipment are used to transport voice,
availability depends on proper functioning of such equipment. (see PARS Network Services agreement).
(2) Access to premises by vendor personnel.
(3) For voice mail networking, all equipment must be at least
VPIM (voice profile for Internet mail) compatible.
D. Service Standards
(1) 99.97 percent availability
(2) Trunking P-.001 (Erlanger C – maximum 0.1 percent blocking
of outgoing calls)
(3) Installation 30 to 60 days from notification, depending on
size and complexity of applications/functions required, as
well as number of users.

SPECIFICS

1. Performance Measures
   A. VCG will provide daily exception reports (via intranet) which display trouble calls by category, and relate to customer agreements.
   B. VCG will provide calling summary reports monthly.
2. Problem Management and Notification Procedures
   A. Service Complaints and Unplanned Outages
   (1) Customer
   (2) Report complaints or difficulties to VCG Help Desk as soon
as possible.
   (3) Be prepared to provide basic information to VCG Help Desk
as described below.
   B. Call Center, Department #41
   (1) Single point of contact for problem reporting by the customer and problem status updates from VCG or other assignee.
   (2) Track problems via Phaedra Tracking System and assign
problem priorities of 1, 2, or 3 with customer concurrence.
   (3) Establish a full description of the problem, including:
   a. Who is affected by site, building, floor, and room?
   b. What symptoms do the customers observe?
   c. What is the known impact of the problem on the affected
customers?
   (4) Provide customers the earliest best guess of the planned resolution to the problem, and estimated time to correct, consistent with the schedules in the priority guidelines.

C. VCG
    (1) Provide Call Center, Department #41 (CCD41) designate with all available information and provide updates consistent with priority guidelines.
        a. Priority 1 — every 45 minutes
        b. Priority 2 — every 4 hours
        c. Priority 3 — once daily
    (2) Provide CCD41 designate with planned resolution and estimated time to fix.
    (3) When the fix is complete, inform CCD41 designate how the fix was tested and verified.
    (4) Provide CCD41 designate with a contact should the problem reappear.

3. Notification Procedures for Planned Outages and Maintenance
    A. Voice Communications Services
        (1) VCG will be the main point of contact for voice communications services.
        (2) VCG will notify customer contacts of scheduled outages no less than 24 hours before or as early as possible.
        (3) The following information will be provided:
            a. A full description of the planned outage
            b. Who will be affected? — specific site, building, floor, room number.
            c. What is the expected effect of the outage, and what changes will the customer observe?
            d. What systems or applications will be affected?
            e. What is the overall work plan?
            f. Will any customer resources be involved? Who?
            g. What is the estimated time to complete the project?
            h. Who is responsible for managing the project and how can that person be reached?

        If the actual work time is expected to exceed the initial projection, CCD41 designate will be notified of the status. CCD41 designate will coordinate with customer contacts and VCG to determine whether the backout plan will be exercised, or the work time will be extended.

4. Escalation Procedure
    A. Voice communications services
        (1) Initial trouble report is made to VCG Help Desk.
        (2) Contact the VCG Help Desk prior to any escalation (765-455-8876).
        (3) The decision to escalate should be made on a case by case basis, by the customer, based on the following conditions:
            a. A status/update call has not been sent by VCG

b. CCD41 priority service
c. Recurring problem that has become chronic
(4) If after the initial trouble report, the customer is not satisfied and chooses to escalate, the following time frames are suggested as a guide. It should be noted, however, that the customer could always contact VCG whenever they desire. These schedules are only provided to assist the customer and to allow the initial maintenance process to be productive.
a. Normal Out of Service Condition
   (i)   Escalate to Manager   2 hours
   (ii) Escalate to Director   4 hours
   (iii) Escalate to Vice President 8 hours
b. CCD41 Priority Service
   (i)   Escalate to Manager   1 hour
   (ii) Escalate to Director   2 hours
   (iii) Escalate to Vice President 4 hours
c. Recurring Problem
   (i)   Escalate to Manager   Immediately
   (ii) Escalate to Director   1 hour
   (iii) Escalate to Vice President 2 hours
(5) Should the customer need to contact VCG for escalation reasons or other information, the following list should be used:
a. Level 1: Jane Sappho   Manager, VCG (345) 789-0989
b. Level 2: James Smyth Manager, VCG (345) 789-6800
c. Level 3: Bill Scopas   Director, VCG (345) 789-2819
d. Level 4: Patricia Pericles VP, PARS Corp. (345) 789-1295

5. Renegotiation Considerations
   A. CCD41 will review this Service-Level Agreement quarterly and apprise VCG of any changes it would like to request.
   B. VCG will review this Service-Level Agreement semi-annually and apprise CCD41 of any changes it would like to request.
   C. Changes must be mutually agreed upon.
   D. All modifications must be made and agreed to in writing.
   E. Either party may request a review of this Service-Level Agreement at any time.
6. Factors that affect renewal/termination of the SLA:
   A. Excessive costs in relation to benefits
   B. If VCG fails to provide services as stated in this Service-Level Agreement
   C. If CCD41 fails to perform tasks as stated in this Service-Level Agreement

Appendix D
CTI Success Stories

THE FOLLOWING CTI SUCCESS STORIES DEMONSTRATE THE DRAMATIC SAVINGS AND SERVICE-LEVEL IMPROVEMENTS that a well-designed CTI implementation can provide to the organization. This material is provided courtesy of Apropos Technology (www.tdata.com).

COMARK, INC.

Using call center management software from Apropos Technology, Inc., one of the nation's largest computer resellers quadruples business in five years, but only has to add four more customer service agents. Cutting agent stress is a welcome bonus.

INTRODUCTION

Some 20 years ago, two fraternity brothers met over beers and started a value-added reseller business called Communications Marketing, eventually shortened to Comark, Inc. Now, with 950 employees and 1997 sales topping $1 billion, Comark and its affiliates provide national and regional distribution, sales, and service of business computer hardware, software, and peripherals for over 500 of the nation's leading manufacturers, including IBM, Hewlett–Packard, Compaq, Toshiba, and Microsoft.

That is a lot of orders and complicated logistics. Business users, pressured from above by their bosses, tend to become emotional about ship dates, returns, damage, and other customer service matters. This adds up to a high-stress customer service environment (not to be confused with technical support, which operates separately).

Holly S. Dobos is Comark's customer service manager. Every day, she and her staff of 15 service agents field 800 and sometimes as many as 1200 calls from harried customers. "Customer service tends to be a call dumping ground," she says. "We get everything from A to Z. Or we used to until we got Apropos."

Problem

Before the arrival of Apropos, everything — technical questions on the hardware and software, catalog orders, etc. — funneled into customer service.

Spanish-speaking customers were routed around until someone could be found to speak with them. Reports had to be requested from "the phone guy," as Dobos refers to the employee who ran the automatic call distribution (ACD) system that formed the primitive backbone of the company's former call handling system. "Those reports were so cryptic," Dobos says, "they might as well have gone straight into the trash."

More importantly, callers sat on hold for three to five minutes. About eight percent of those callers gave up and hung up. When the service rep did pick up, at least a minute of what Dobos calls "garbage" ensued — name, purpose of call, resolution of past calls, etc.

According to Dobos, customers do not want to be handled by someone who has to ask a lot of elementary information and whom they think cannot handle their problem. Under the old system, customers would often ask for a supervisor right at the outset. "Every customer wants to think their problem is special and they are getting specialized treatment," she says. "I almost had to have two supervisors, one to supervise and the other to play the supervisor on the phone all day."

Solution

There had to be a better way. The two company owners, Dobos says, found one. They called in Apropos Technology, listened to a presentation, and became a beta test site for Apropos. "We are in the business of selling computers," Dobos says. "Our owners are very cutting edge."

Consisting of four components — the Telephony Server, the Agent Desktop, the Supervisor Desktop, and the Report System Desktop — Apropos runs on a Compaq Apropos Server, front-ended by an IVR (interactive voice response) system, which allows some customers (about 25 percent) to get their problems solved without talking to an agent (order status, addresses, etc.). Then, Apropos uses customer identifiers (phone number, serial number, or the like) to route past customer history to the agent's desktop in the form of screen pops, so agents know at a glance what they are dealing with when a call comes in.

Benefits

"I cannot emphasize enough the importance of dealing with the emotional side of CTI (computer-telephony interface)," Dobos says. "Knowledge is power/control. With Apropos, the agent leads the call with knowledge. He or she knows who is calling and why. That's a big thing for us. When you take a call and say, 'How can I help you?,' you've left yourself open for anything the customer wants to dump on you. You're really flapping in the wind."

"Call centers are stressful enough," Dobos continues. "Agents need control over their own desktop. With Apropos, they can see how many calls are there, and how fast they are dropping in. The agent can pick up the phone and say, 'Hi, Mr. Jones, are you calling about your printer?' That way, you've defused your customer instead of letting the customer take you down the path, which could be a couple of miles before you get back control. This makes a huge difference for us."

Comark empowered its agents by turning over the basic design of the system to those who would be using it. "Our customer service group decided on the prompts. Our people even recorded some of them, in English and Spanish," Dobos recalls. "That was truly their decision, and they felt strongly about doing it. A lot of managers might be afraid of that. In a 'take' environment in which agents can opt to 'take' calls, would they refuse certain calls? Not here. They 'own' this system and they fight to take calls off the queue."

"I can't emphasize this enough," Dobos says. "When we can give the agents control over the call, they can pick up the phone without wincing, so they take a lot more calls. This gives stability to the call center. Most call centers have a high turnover. We don't."

Apropos routes calls more efficiently. "It streamlined the system," Dobos confirms. "Tech support calls go directly to tech support. Catalog orders go directly to the catalog area. Overnight, we became customer service agents rather than telephone operators."

Spanish-speaking customers show up only in the queue of a Spanish-speaking agent. "We also can move people in and out of different call groups," Dobos says. "That way, people who need a specialist for a certain transaction can work with that person. Your job is less stressful if you handle the calls you are qualified to handle." This applies particularly to new hires, according to Dobos. "We can tell a new person, 'You're only going to be handling order status calls at first.' Then we can add a function each day. This is also good for seasonal hires."

Hold times have gone down from three to five minutes, to 40 seconds or less. The abandon rate plummeted from 8 percent to 3 percent, with the added ability to call back those abandoned calls and rescue some of them. Calls are shorter, too. Where it used to take one minute to get to the meat of a call, with calls lasting from 5 to 15 minutes, now a five-minute call is a long call, Dobos says. "We cut the talk in half," she says.

The phone bill also fell, with 800 charges reduced as much as 25 percent. "Of course, we have to take some extra time to brag," Dobos smiles. "People want to know how we knew their name and problem. They can't believe it. It's very motivating to brag about what you do. It makes the agents want to take more calls."

In all, business has quadrupled and the call agents have gone from 11 to 15, only a 35 percent increase. Each agent can take three to five percent more calls than they could with the old ACD system.

Dobos also praises the Apropos Wrapup window, allowing customer information to be written to the mainframe for retrieval the next time the customer calls. "I designed the window myself to do what I wanted," she says. The same goes for her new report generating capability. Instead of the "trash" reports churned out by the ACD system, she gets what she wants when she wants it. "I don't need a 25K-sized report to know what's going on," she says. "I already know what's going on."

"I also like being able to turn off the queue from the desktop," Dobos says. "The first people to arrive in the morning turn it on, the last out at night turn it off. Before, I couldn't have a staff meeting; the calls kept coming. Now, with Apropos, I can pick up the phone and record a message saying we will be closed for an hour. The Message of the Day feature is great."

What about tech support from Apropos? "When we got the system, they came in and worked all night," Dobos says. "The next morning, my agents turned it on. That was that."

Summary

An early tester and adopter of the Apropos call management system, Comark, Inc., a $1 billion value-added reseller of computers and peripherals, reduces agents stress, call length, phone bills, and agent turnover.

Key Benefits
- Apropos call management system cuts agent stress by allowing agents to use customer history information to gain control over the caller and their desk, and in the process, shorten calls.
- Apropos therefore has allowed the company to quadruple in head count, while the call center has grown only 35 percent.
- Hold times have dropped from three to five minutes, to 40 seconds.
- Call times have been reduced 40 percent.
- Abandoned calls have been cut from eight percent to three percent.
- Supervisors can control the queue and turn it off when necessary.
- Calls are routed quickly and accurately, benefiting Spanish-speaking customers, in particular.
- Shorter calls make for lower 800 phone bills, even allowing time for agents to explain to amazed and pleased customers how they knew the customer's name and problem without being told.
- Customers feel special and well-cared for, virtually eliminating demands to "speak to a supervisor."

"I can't emphasize this enough. When we can give the agents control over the call, they can pick up the phone without wincing, so they take a lot more calls. This gives stability to the call center. Most call centers have a high turnover. We don't."

— Holly S. Dobos, Customer Service Manager, Comark, Inc.,
Bloomingdale, IL

Systems At a Glance

- Compaq server (Apropos Technology, Inc.)
- software
- OS/2, converting to Windows NT
- SAP coming as client/server application suite
- terminal emulation to get data from the AS/400
- networking
- Ethernet

3COM CORPORATION (FORMERLY U.S. ROBOTICS)

Using Apropos running on an NT platform to manage its customer calls, 3Com saves more than $6 million a year in headcount costs; increases customer and agent satisfaction; and captures customer and product data that has led to timely product enhancements fueling the company's phenomenal growth.

As the leading manufacturer of the modems fueling the Internet age, U.S. Robotics, now 3Com Corporation's client access division, also manufactures, sells, and supports remote access servers, enterprise communications systems, and desktop/mobile products, all of which enable customers to manage and share data, fax, and voice information. When this communications giant needed an advanced communications system of its own, who got the call? Apropos Technology.

"Some companies turn to a new call distribution system because their agents are overloaded. Not us," says Lael Miller, support programs manager for 3Com, in Skokie, IL. "Instead, we looked at the situation and said, 'We're a high tech company and we have to go with the best technology.'"

Like any company selling a wide line of complicated products, 3Com gets over three million calls a year. In 1995, 60 agents were handling up to 1000 calls a day, using a Rolm ACD (automatic call distribution) system with Octel Voice Mail to provide the auto-attendant function and prerecorded tech support messages answering "most-asked questions." The average customer waited 12 minutes for expert advice. How many customers got tired of waiting and hung up? That was hard to tell. Capturing information such as call abandonment rates was not the strong point of the Rolm/Octel system.

"We needed a system that was just as reliable, but more flexible, and one which would provide better and more accurate statistics as well as real-time information," Miller adds. Useful realtime information would include who was on hold, what each customer wanted, and which agents were talking to whom. The reports generated by the company's ACD system needed to be improved to provide management with the information needed to distribute the workload properly. In addition, the old ACD system would not map agent skill sets, allowing agents to be put in line for all the calls they were qualified to answer. An expert in the Sportster modem, for example, might get only Sportster calls, when he or she was also qualified to handle Courier modem calls.

Those changes were not all 3Com wanted in a new system. The wish list was lengthy. The new process also had to work with any PBX switch, and be compatible with OS2/Windows, Win 95, and NT operating systems. It had to incorporate ANI (automatic number identification), DNIS (dialed number information service), and touch-tone services from phone carriers. The company also wanted a visual queue, allowing agents and supervisors to pick and choose between incoming callers. Apropos offered that.

Other major requirements appearing on the 3Com wish list included: interactive voice response, auto attendant, screen pops with customer data, holdtime announcements, live drag-and-drop agents, voice mail, abandoned call capture, closed call capture, work@home for remote agents, and hooks to email so agents could fax or e-mail promised material instantly. "We shopped this list to about 40 companies," recalls Miller. "Apropos Technology was the only company that could provide 100 percent of this list. The second-closest could only offer 70 percent and cost six times as much."

Apropos interfaces with the company's existing network databases and telephone switches. With Apropos, when 3Com customers call, they are prompted for an identifier, such as a warranty number, telephone number, or repair number. The first service they are offered is the interactive voice response (IVR) portion of the system, providing 24-hour, seven-day-a-week access to recorded status reports on their repair orders. Without accessing a live agent, customers also can determine what files to download to upgrade their particular software. "Forty percent of the callers," Miller says, "get their answers there and don't need to speak to an agent. This is a remarkable percentage."

If the IVR system cannot meet a customer's needs and the call continues to a live agent, the customer's identifier is passed to the 3Com network server and the customer's record is instantly accessible to the agent in the form of a screen pop before the agent even lifts the receiver. "You can see what the customer wants before you even say hello," Miller says. "The agent can pick up and say, 'Hi, John, this is Marty, are you still having a problem setting switches?'" This is where speed or time savings really come into play," Miller says. "We can save as much as 45 seconds a call," he estimates, "because we don't have to ask who it is or why they are calling." Although most agents are qualified to answer questions on more than one product, each has certain skill areas. The "caller preview" function (or Q-view) allows agents to preview caller information. That way, the most qualified agent can pick up and assist the customer. Agents know at a glance how many are in their queue and how long each has been waiting, and supervisors have direct visibility into the call center at all times. They can get the complete picture in real time: how many calls are in the queue, being talked, dropped, resolved, or being resolved. According to Miller, customers, management, and agents all benefit from Apropos. Notably, 40 percent of the callers can be helped through the IVR system and do not have to wait for a live agent. "This helps individuals and businesses get support 24 hours a day, seven days a week," Miller explains. When customers must speak to an agent, they have to wait less time and need not repeat their identifiers and problem over and over. "Not only does this save an enormous amount of time (45 seconds on the average eight-minute call), but one of the best ways it helps is on repeat or escalated calls."

Although Apropos has slashed customer wait time from 12 minutes to two minutes (an almost 85 percent improvement), 3Com, like every call center, experiences some call abandonment. "About 90 percent of those hang up less than a minute after being put on hold, so they were going to hang up anyway," observes Miller. "Apropos, however, allows us to capture the numbers of everyone calling in, so we can call those back who, for some reason, were kept waiting a length of time. This improves customer service. We have a customer satisfaction team that calls people; we do mailings, too," he adds. "We know our customers appreciate this system."

From management's viewpoint, the advantages of Apropos are many. Number one is the increased speed of call handling, because speed translates to a need to hire fewer agents. "We went from 1000 to 12,000 calls a day since 1995 — a 1200 percent increase," Miller says, "but we only needed to increase the number of our agents by 360 percent. That way, we save over $6 million a year in headcount costs."

"We're in growth mode," Miller says. "Another good thing about Apropos is that it is readily expandable. We've been able to change direction rapidly, push the envelope. We went from 48 ports, to 72, to 225, from one server to six. Apropos is modular; you can add as much as you want. The NT platform is really cool."

Of Apropos Technology, Miller adds: "There have been times when I've run into a brick wall with other companies, over software leasing and so on, but with Apropos if I need software written, it's on my desk the next morning. No matter how fast we grow, we know Apropos Technology can keep up. We have no idea what volume we'll be facing on this…well, I'd call it a rollercoaster, but it only goes up, never down."

Key Benefits for 3Com

- Apropos Interactive Voice Response (IVR) component allows 40 percent of its inbound callers to track repairs and get commonly answered questions answered automatically, without the use of a live agent.
- Apropos saves 45 seconds on the average eight-minute call, requiring the hiring of fewer agents. Since 1995, this has resulted in an estimated savings of $6 million in agent headcount costs.
- Average wait time has been cut from 12 minutes to two minutes, an 85 percent improvement.
- Apropos visual queue allows agents to preview calls and always routes calls to the best available agent, thus increasing both agent and customer satisfaction.
- Skill-based routing allows agents with multiple skill sets to sit in multiple queues and accept all appropriate calls.
- Customers do not have to repeat information, and having a customer's record at their fingertips saves agents hundreds of keystrokes and repetitions on repeat calls and escalations.
- Supervisors can monitor the call center from their desktops, including call load, call disposition, and the activity of individual agents, and can allocate agents between queues.
- Geographically separated call center sites operate off the same database but different telephone switches, saving money and space, yet all are managed from the Chicago location. Agents even can work at home.
- Call and product data captured by the system can be analyzed to improve 3Com operations and product base.

WORLDWIDE ACCESS/VERIO

Relying on Apropos Technology, Inc., to run its call center, pioneer Internet access provider cuts administrative costs in half and scores 90 percent customer satisfaction from its 10,000 demanding customers.

Introduction and Background

One of the first companies to offer Web site hosting, WorldWide Access/Verio provides not only Internet access and Web site development, but dedicated lease lines and firewalls as well. Several years ago, when the Internet was the global equivalent of the telephone party line, WorldWide Access/Verio could field calls on a couple of Centrex phone lines. But now, as the Chicago-based company has shown, helping people use today's Internet requires up-to-date call handling.

Enter Apropos Technology, whose Apropos call management system helps the four-year-old Internet company stay on top of the questions and needs of its 10,000 technology-savvy customers. Referring to his fast-growing client base, Gregory Gulik, executive vice president of WorldWide Access/Verio, says, "In no time, we had more calls, needed more people, and everyone didn't know the answers to everything anymore. That's where Apropos entered the picture."

Problem

In technology years, three is a pretty large number. Not only has its client base grown, but the Internet services and applications WorldWide Access/Verio's customers request have grown as well. WorldWide Access/Verio prides itself on its customer service — an average call lasts 15 minutes as the techs explain the company's services and try to get the best fit between customer and product. Three years ago, customers calling about a technical problem, billing question, or support issue encountered one of only six operators. Another obstacle to consistent customer service was the fact that the access window was only eight hours. At night, exasperated or inexperienced users got an answering machine, which did little to defuse their frustration.

Solution

"We looked at several call handling systems," Gulik recalls. "Apropos was the most cost-effective. It was standards-based, would work with our old system, and would allow us to add features of our own. Apropos, to put it simply, would do anything we wanted it to."

WorldWide Access/Verio utilizes the entire Apropos system, which consists of four major components: the Apropos Server (the gateway between the call center and the public network), the Agent Desktop (the link between the customer on the line and the call center), the Supervisor Desktop (the link between the agents, call managers, and supervisors), and the Report System (a full-featured reporting system documenting each call and overall operations).

Operating 24 hours a day, Apropos acts as a call attendant, meaning no more lost calls or busy signals. Using Apropos' queuing and distribution capabilities, WorldWide Access/Verio quickly routes callers to the proper department. Its audio text feature allows callers to obtain information on products and services without speaking to an agent, which in turn frees up agents to solve more complicated problems. Hold times are now ten seconds or less. At WorldWide Access/Verio, the call center consists of only 16 agent stations. Yet, even with hundreds of calls a day, because of Apropos, Gulik says he cannot remember a time when there were more than two or three people on hold.

Of course, he knows that if more than two or three people stacked up on hold, the Supervisor Desktop would immediately notify the call center manager. The Supervisor Desktop allows the call center manager to see at a glance who is working, what the hold times are, how many calls an agent has answered, and how many are in each queue. "This is an excellent real-time view of our call center's operations," Gulik remarks.

Apropos flexibility allows WorldWide Access/Verio to run multiple applications, such as allowing agents to verify the status of the computer system at a glance. This is vital when calls involve lost e-mail or inability to access the Net. "The agent knows immediately if the problem is coming from our end," Gulik says. "Knowing the status of our system, the agent can say, 'Oh, yes, our system will be down for ten minutes,' or whatever. The customer knows exactly what is going on immediately. Conversely, if there is no system problem, the agent can start checking potential problems at the customer's end. Is your modem properly configured, and so on. We don't have to put customers on hold to see what the problem is."

Gulik was pleased to learn that the interconnectivity features of Apropos allow the company's call agent to run from remote sites. "People with two phone lines can work at home," Gulik says. "One phone is logged on to Apropos over the Internet, the other is for answering customer problems routed from the office." This means that on days when call volume is high, extra agents can work from home.

At WorldWide Access/Verio's call center, agents work from a script when dealing with customers, asking such questions as, "Is your Caps Lock turned off?" "Are you using the correct password?" As the agent works with the customer, he or she enters notes into the database, such as the name of the agent the person talked to, what the agent recommended to the customer, a diagnosis of the situation, etc. That way, if the user calls back, the agent who answers knows what has been tried and what has not, as well as other relevant account history. This reduces repeated explanations, prevents customer frustration, and cuts calling time.

Benefits

Greater customer satisfaction is one of the benefits of the system. "Our major motive in getting Apropos," Gulik says, "was speed and quality of response to our customers. With Apropos, the customer gets to the person who knows how to deal with the problem. Customers aren't told five times to try something that doesn't work."

As part of their business practices, the company regularly does customer satisfaction surveys. "They have been phenomenal," Gulik says. "Greater than 90 percent satisfaction. People really like to get an answer, get an explanation, and know what's going on."

Another major benefit is the functionality of the Supervisor Desktop, Gulik says, which gives call center managers realtime visibility into the center's operations, allowing them to check activities such as agent performance, group performance, and call load. "We can keep an eye on things. We know who's on the line, how long they have been on, who is slowing down, and if we need more people." He notes that last one with a smile. Because they have more information on the caller and problem and can respond more effectively, agents are less frustrated, thus improving job satisfaction. Agent turnover at WorldWide Access/Verio is low. "Only three people have left since we started four years ago," Gulik says. "With Apropos, our agents are less hassled."

Overall, Gulik estimates, the company has realized 50 percent savings in operating costs due to its more efficient use of its people, including the ability to work from a remote site. The first year, that amounted to $450,000 back in the company's pocket.

As WorldWide Access/Verio heads toward the future, it anticipates that Apropos will continue to meet its needs. Future plans call for a transition in the database to Lotus Notes and the addition of Centrex Caller ID to populate the Apropos system. Because Apropos can work with Caller ID to enhance the Agent Desktop, this will expand WorldWide Access/Verio's capabilities even further. Marketing efforts will also be refined. Separate phone lines will be keyed to various company ads, allowing tracking of which ads pulled better than others. Gulik has no doubt that Apropos can continue to grow with the company. "With Apropos," he says, "anything we think would be cool to do, we can."

Summary

WorldWide Access/Verio, an Internet provider and early Web site host company with 10,000 customers, uses Apropos to bring almost instant responses to user questions, resulting in a greater than 90 percent customer satisfaction rate, a 50 percent drop in operating costs, more targeted use of personnel, and more efficient supervision of its call center.

Key Benefits

- Apropos flexibility means that WWA/V can use Centrex now and still shift to another switch in the future.
- Inbound customers get a response; no more busy signals.
- Customers are routed to the person most likely to have the answer they seek, which has resulted in greater than 90 percent customer satisfaction scores.
- The responding agent also will know what has gone before, so the customer does not have to repeat.

- By capitalizing on Apropos' ability to interface with multiple applications, agents can assess the status of the company's computer system and know if the caller's problem is an internal computer error or something the customer is doing incorrectly.
- Call center supervisors can get instant realtime views of center operations, including who is on the line, how many calls they have taken, have waiting, etc.
- Agents can, on occasion, link to the call center and work from home.

"Apropos works with our old system, and allows us to add features of our own. Apropos, to put it simply, does anything we want it to."

— Gregory Gulik, executive vice president, WorldWide Access/Verio

Systems at a Glance

- hardware
- Apropos Server
- Pentium IBM-compatible PCs with Windows 95 and NT
- Centrex telephone switch
- PBX (future)
- software
- Apropos Agent Desktop (Apropos Technology)
- Apropos Supervisor Desktop (Apropos Technology)
- Apropos Report System (Apropos Technology)
- Lotus Notes (future)

SEAGATE SOFTWARE, INC.

Information Management Group

Using the Apropos total interaction management system, the Information Management Group of this business intelligence software developer pays for the entire system in 12 months, improves customer service, uses the Internet to connect four sites, and reduces call agent stress...Really.

Seagate Software, Inc., of Vancouver, B.C., Canada, a subsidiary of Seagate Technology, Inc., in Scotts Valley, CA, develops business intelligence tools and applications. In strategic relationships with such companies as Microsoft, Novell, and IBM, Seagate Software shipped three million copies of Seagate Crystal Reports. When using the company's products, customers rely on timely and accurate technical support.

Often, customer inquiries are complex, protracted, and concern large business information systems. Until a year ago, customers calling the Information Management Group of Seagate Software encountered a six-queue ACD system that stacked the calls until an appropriate technical support representative became available. To direct people to the right queue, the company employed as many as 11 "pre-call processors" (PCPs). Although qualified as agents and earning agent salaries, these employees were basically acting as "operators" — just answering phones.

"We had a high wait time," admits David Galloway, the company's Director of Technical Support. "It took two to three minutes to get to the PCP, and then an average of 20 minutes, or sometimes as much as an hour, to reach the appropriate support team member for help. Often, callers had to explain the problem twice, which took time. And people really didn't get any information while on hold about what their wait time was or what the process of handling the call would be. Our call abandonment rate was high."

Then CallPro, a Canadian telecommunications software products reseller and integrator, told Seagate about call management software from Apropos Technology, Inc. First, according to Galloway, CallPro "opened his eyes" to agent empowerment. Apropos' ability to display all calls and allow agents to select those calls that took priority, had been waiting longest, or fell in a specialty area. "Knowing what the caller wants in advance helps agents be more psychologically prepared," Galloway adds.

More importantly, the Apropos system substituted a 1.5 minute IVR routine for the costly system of pre-call processing. Callers are even allowed to choose the music and information they want to hear. "We knew we should be able to automate this, and with Apropos we could," Galloway says. "Apropos gave us more capability, flexibility, and greater ease of use for less money [than its competitors]. It had a very fast payback time."

"The biggest, number one issue was how to eliminate the need for the PCPs and move those individuals from operator status to tech support agent," Galloway says. People were being hired as agents and relegated to an operator. "They were not adding a lot of value," Galloway says. "After you do that job for a day, it gets boring. People were getting stuck there." The supervisors, too, were being hampered by the lack of realtime information on the status of the call center. To switch an agent from one queue to another involved reprogramming the PBX system; in other words, putting in a request to IT and scheduling a fix — work that took days or weeks, not seconds. As for assessing incoming calls, the closest thing the company had to realtime information was a periodic recycling of the message reader board. With 14 queues (seven priority and seven regular), this meant 14 message boards downloaded to the network and flashed every minute. "It worked, but it was kludgy," Galloway says. For a company that takes service seriously, the welfare of waiting callers was also an issue. "We really didn't know how long a caller had been waiting," Galloway says. "We could see a wait time on the next caller only.

Both those who had purchased our regular service and those who had contracted for premium service were fed into the queues, with the priority customers being put ahead. The first person might only have been waiting two minutes, and even though we might want to take a caller with regular service who had been waiting 40 minutes next, we didn't know about that person. We couldn't see what was going on, and meanwhile people were sitting in Waitville." Making the wait as pleasant as possible was also a priority. "A 40-minute wait is a long wait," Galloway says. "Some people didn't like the music or wanted no music. We now have five choices for our callers — four kinds of music and silence.

Having compiled a varied wish list of functionality's, Seagate Software looked at a half dozen vendors ranging from low-end voice recognition options to Apropos' competitors. Apropos won. "One significant factor was that Apropos would lower the cost of our PBX switch," Galloway says.

Because of the product's ACD functionality, Apropos allowed Galloway to pay only for a "plain vanilla" PBX connection. The fact that Apropos, through CallPro, offered local support, service, and training, sealed the deal. "We didn't want to wait for someone to come from Chicago, and we didn't have to," Galloway says. "Apropos and CallPro have done a superb job."

When customers call now, they execute a short IVR routine with the option of reaching a "real person" by dialing "O" for operator. Instead of waiting several minutes to tell their problem to a PCP (which, in all likelihood, they would have had to repeat later to the tech support agent), callers answer a couple of questions using the telephone keypad, such as their registration number and type of service for which they are licensed (regular or priority). Prompts allow them to change their answers, if need be. Once in the queue, they are advised of their position in the queue. "Apropos allows for lots of realtime queue information. We found that telling callers their position in the queue worked best for us," Galloway says.

"Apropos has paid for itself in terms of eliminating the PCP function," Galloway says. Callers used to wait a couple of minutes to even talk to one of almost a dozen PCPs; now they wait 45 seconds or less and are handled by one or two people, using IVR. "We would likely have had to hire ten more people if we had not put in Apropos," Galloway says. "We are using our reps more effectively and productively. The system paid for itself in less than a year. And, by moving these employees into tech support, the company was able to cut wait time after the IVR phase of the call."

Galloway also praises the flexibility that Apropos gives his agents. "Agents can take the calls they want, when they are ready," he explains. "The voluntary nature has never been a negative. Some people argue that agent empowerment will mean that agents simply won't pick up; but we've found just the opposite to be true. Being empowered has lowered stress and encouraged agents to handle calls."

Apropos is easy to work with, he adds. "We have had a lot of changes as we've grown. We can move people from one queue to another in a second.

We can also put agents in more than one queue. Try doing that with most ACDs. This allows supervisors more flexibility in making agents with various specialties available to more callers. Other systems that do it are pretty expensive. This is a pretty big deal and most people don't realize how helpful it is." Seagate Software uses the database look-up functions to check entitlement for its 7 × 24 Priority Support Program. If the customer is not registered for 7 × 24 support, they are advised by the system. "They are still serviced," Galloway points out, "but the agent knows that the system told them about the violation so the agent can focus on customer service and not on service policy. We can do that later with the account executives." Any unexpected pluses? "Yes!" Galloway exclaims. "After our purchase, we discovered how well Apropos runs over the Internet. Our offices and outsource partners in Vancouver (B.C.), Florida, California, and England can all look at the queues at the same time and pick up calls. It's pretty amazing." Using the Apropos total interaction management system, the Information Management Group at Seagate Software was able to pay for the entire system in 12 months, improve customer service, use the Internet to connect four sites, and reduce call agent stress…Really.

Key Benefits

- Using Apropos' IVR functionality, Seagate Software was able to move at least ten agents from operator to tech support status, saving the salaries of those positions and eliminating boring jobs.
- The system paid for itself in a year.
- Moving agents from operator to tech support status also shortened wait time because more tech support people were available to take calls.
- Callers can key in their information and then select the music they will hear as they wait for a tech support agent. Position in queue and key informational messages targeted by queue are also announced.
- Apropos allows agents to see all callers at a glance and select those who have been waiting longest or have licensed premium service.
- Supervisors can switch agents from one queue to another in less than a minute or put agents in multiple queues.
- Allowing certain agents to specialize in "trouble" calls has cut the time of those calls in half.

Appendix E
Representative Telephony and Communication Vendors

ACT Networks Inc.

Address:	188 Camino Ruiz
	Camarillo, CA 93012
Tel:	(805) 388-2474
Toll-free:	(800) 367-2281
Fax:	(805) 388-3504
E-mail:	info@acti.com
Web site:	http://www.acti.com

Apex Voice Communications Inc.

Address:	15250 Ventura Blvd., 3rd Floor
	Sherman Oaks, CA 91403
Tel:	(818) 379-8400
Fax:	(818) 379-8410
Web site:	www.apexvoice.com

Apropos Technology

Address:	3400 188th St., SW, Ste. 185
	Lynnwood, WA 98037
Toll-free:	(800) 729-4767 (800-729-4POS)
Web site:	http://www.aproposretail.com/

AT&T

Address:	32 Avenue of the Americas
	New York, NY 10013
Tel:	(212) 387-5400
Web site:	http://www.att.com/

BCS Technologies
Address:	1950 Old Gallows Road, Ste. 201
	Vienna, VA 22182
Toll-free:	(888) 790-8227
Fax:	(703) 790-0098
Web site:	http://www.bcs-online.com/

Boole & Babbage, Inc.
Address:	3131 Zanker Road
	San Jose, CA 95134
Tel:	(408) 526-3417
Toll-free:	(800) 544-2152
E-mail:	aknapp@boole.com
Web site:	http://www.boole.com/

Brooktrout Technology
Address:	410 First Avenue
	Needham, MA 02494-2722
Tel:	(781) 449-4100
Fax:	(781) 449-9009
Web site:	http://www.brooktrout.com/contact_us/index.html

CCOM Information Systems
Address:	120 Wood Avenue South
	Iselin, NJ 08830
Tel:	(732) 603-7750
Fax:	(732) 603-7751
E-mail:	info@ccom-infosys.com
Web site:	http://www.ccom-infosys.com/fdbk.htm

Centigram Communications Corporation
Address:	91 E. Tasman Drive
	San Jose, CA 95134
Tel:	(408) 944-0250
Fax:	(408) 428-3732
Web site:	http://www.centigram.com

Cisco Systems, Inc.
Address:	170 W. Tasman Drive
	San Jose, CA 95134
Tel:	(408) 526-4000
Web site:	http://www.cisco.com/

Décisif Software Solutions Inc.
Address:	4 Place du Commerce, Ste. 570
	Montreal, Quebec H3E 1J4 Canada

```
      Tel:   (514) 362-7117
Toll-free:   (888) 517-2929
      Fax:   (514) 362-0456
   E-mail:   info@decisif.com
 Web site:   http://www.decisif.com
```

Eyretel *Ltd.*
```
  Address:   2 Horsham Gates
             Horsham, West Sussex
             RH13 5PJ United Kingdom
      Tel:   +44 (1) 403-214-400
      Fax:   +44 (1) 403-214-420
 Web site:   http://www.eyretel.com
```

Infointeractive
```
  Address:   1550 Bedford Hwy., Ste. 604
             Bedford, Nova Scotia B4A 1E6 Canada
      Tel:   (902) 832-1014
      Fax:   (902) 832-1015
   E-mail:   info@infointeractive.com
 Web site:   http://www.interactive.ca/
```

Inova Corporation
```
  Address:   INOVA's Main Office is located in Charlottesville, VA
      Tel:   (804) 817-8000
      Fax:   (804) 817-8002
 Web site:   http://www.inovacorp.com/
```

Invade Corporation
```
  Address:   20 Apex Court
             Almondsbury, Bristol
             BS32 4JT United Kingdom
   E-mail:   info@InVADE.net
             sales@InVADE.net
             marketing@InVADE.net
      Tel:   +44 (0) 171 575 0048
      Fax:   +44 (0) 171 575 5948
 Web site:   http://www.invade.net/
```

ISI Infortext
```
  Address:   1051 Perimeter Drive
             Schaumburg, IL 60173
      Tel:   (847) 995-0002
Toll-free:   (800) 366-6550
      Fax:   (847) 995-0003
 Web site:   http://www.isi-info.com/
```

Lucent Technologies
    Address: 600 Mountain Avenue
                Murray Hill, NJ 07974-0636
    Web site: www.lucent.com

NEC Corp.
    Address: 8 Corporate Center Drive
                Melville, NY 11747
    Tel: (516) 753-7000
    Fax: (516) 753-7041/7042/7043
    Web site: www.nec.com

Nortel Networks
    Address: Attn: Customer Development
                4401 Great America Parkway
                Santa Clara, CA 95052
    Tel: (800) 466-7835 (800-4-NORTEL)
    Web site: http://www.nortelnetworks.com

Nuance Communications
    Address: 1380 Willow Road
                Menlo Park, CA 94025
    Tel: (650) 847-0000
    Fax: (650) 847-7979
    Web site: http://www.nuance.com/

Parlance Corporation
    Address: 85 Lakewood Road
                Manasquan, NJ 08736
    Toll-free: 1 (888) 700-6263 (888-700-NAME)
    Web site: http://www.nameconnector.com

Pronexus, Inc.
    Address: 260 Terence Matthews Crescent
                Kanata, Ontario K2M 2C7 CANADA
    Tel: (613) 271-8989
    Fax: (613) 271-8388
    E-mail: info@pronexus.com
    Web site: www.pronexus.com

Schmeltzer, Aptaker & Shepard, PC
    Address: The Watergate
                2600 Virginia Avenue, NW, Ste. 1000
                Washington, D.C. 20037-1905
    Tel: (202) 333-8800

E-mail: sas@saslaw.com
Web site: http://www.idsonline.com/saslaw/

Selsius Systems
Address: 18581 North Dallas Parkway
Dallas, TX 75287
Toll-free: (800) 946-2483
Web site: http://www.selsius.com/Default.htm

Southwestern Bell
Address: SBC Communications, Inc.
175 E. Houston
P.O. Box 2933
San Antonio, TX 78299-2933
Tel: (210) 821-4105
Web site: http://www.swc.com

Spanlink Communications, Inc.
Address: 7125 Northland Terrace
Minneapolis, MN 55428
Tel: (612) 971-2000
Web site: http://www.spanlink.com/

Telco Research Corporation
Address: 616 Marriott Drive
Nashville, TN 37214
Tel: (615) 872-9000 (615-TRC-9000)
Toll-free: (800) 488-3526 (800-48-TELCO)
Fax: (615) 231-6144
Web site: http://www.telcores.com/

T-NETIX, Inc.
Address: 67 Inverness Drive East
Englewood, CO 80112
Tel: (303) 790-9111
Toll-free: (800) 531-4245
Fax: (303) 790-9540
Web site: http://www.t-netix.com/

Veritel Corporation
Address: 70 W. Madison, Ste. 710
Chicago, IL 60602
Tel: (312) 803-5000
Toll-free: (888) 837-4835 (888-VERITEL)
Fax: (312) 803-3311
Web site: http://www.veritelcorp.com

Verivoice, Inc.
Address:	5 Vaughn Drive
	Princeton, NJ 08540
Tel:	(609) 924-3000
Web site:	http://www.verivoice.com/

Vocaltec
Address:	35 Industrial Parkway
	Northvale, NJ 07647
Tel:	(201) 768-9400
E-mail:	info@vocaltec.com
Web site:	http://www.vocaltec.com/

Appendix F
Additional Telecommunications Web Sites

These Web sites are in addition to those listed in Appendix E.

http://chatsubo.java.sun.com/products/java-media/jmf/
Java media framework page.

http://china.si.umich.edu/telecom/telecom-info.html
Comprehensive telecommunications site with more than 7000 links.

http://csrc.ncsl.nist.gov/
National Institute of Standards and Technology. Security topics.

http://hyperarchive.lcs.mit.edu/telecom-archives/
Telecom Digest.

http://www.alliancedatacom.com/frame-relay-voice-integration.htm
Voice over Frame Relay.

http://www.altigen.com/
Server-based telephones.

http://www.ambuscom.com/index.html
Telecommunications consulting services.

http://www.ameagle.com/whitepapers/univ_srv_msg.htm
Universal messaging services.

http://www.angustel.ca/Welcome.html
Telecommunications management magazine.

http://www.artisoft.com/telephony/
For developers: Visual voice 5.0; Teladvantage is their "un-PBX" product.

http://www.computertelephony.org/
The Computer Telephony Portal; references to a large number of telecommunications sites, by category.

http://www.computertelephony.org/whatsgoingon.html
Summary of what is going on in the CTI industry.

http://www.critical-angle.com/ldapworld/ldapfaq.html
LDAP Q&A.

http://www.ctimag.com/
Various CTI magazine Web sites.

http://www.emtechnologies.com
General telecommunications consulting services, including Contingency planning and disaster recovery Equipment acquisition, System moves, Project management, Carrier selection, Telephone bill audits, etc.

http://www.iptelephony.org/
Internet telephony resources and white papers.

http://www.kauffmangroup.com/
Enhanced faxing information (over IP).

http://www.microsoft.com/com/activex.asp
Microsoft ActiveX controls.

http://www.netspeak.com/
IP telephony products.

http://www.onepassinfo.com/optelcom.html
Beginner's guide to telecommunications and more advanced topics (tutorials).

http://www.shelcad.com/
CTI products for Internet and intranet telephony market.

http://www.sun.com/products-n-solutions/sw/SunXTL/index.html
The telephony and call management platform for the Sun SPARC/Solaris environment.

http://www.telecomlibrary.com/
Computer Telephony magazine.

http://www.vmen.net/
Internet messaging and voice mail.

http://www.vtg.com/
Voice Technologies Group. Various IP and CT products.

http://www-fr.cisco.com/warp/public/788/7.html
Cisco voice implementation guide.

www.phonezone.com
Tutorials, white papers on a broad range of telephony topics. Excellent site for background information.

About the Author

WILLIAM A. YARBERRY, JR, CPA, IS A TELECOMMUNICATIONS CONSULT-
ant based in Houston, TX. His experience includes telecommunications
management, IT risk assessment, applications systems development, and
IT outsource contract management. He has published over a dozen articles
in such publications as *EDPACS, Information Strategy: The Executive's Jour-
na*l, and *The Internal Auditor.*

Index

PBX (private branch exchange)
architecture, limits of, 151
cable infrastructure of, 172–174
call accounting and, 164–170
circuit connections to CO, 8–9
circuit-switched, functions of, 17
class-of-service parameters for,
180–182
defined, 1–2
and enterprisewide dialing plans,
187
installation of
adjunct processing, 138
backout plan for, 141
class-of-service for, 136–137
current environment survey,
131–134
cutover plan for, 140–141
dial plans for, 136
end-user training for, 135–136
equipment, pre-installation,
137
help desk and implementation,
138–139
implementation committee,
130–131
preparedness review for, 139
project team, implementation,
129–131
routing tables for, 137
software installation, 137–138
station reviews of, 134–136
and IP gateway, 144, 145, 146
vs. LAN, 150–152
and on-net calls, 192
routing tables for, 2
and toll fraud, 179, 180
toll fraud insurance, 182
UNIX operating systems and, 16
PC (personal computer), as
telephones, 57, 58, 61
PCM (pulse code modulation), 10, 12
PC-to-PC voice, 146
Personalized ringing, 20
Personal Station Access, 253
Phoneline applications, 62
Phonemail voice mail application, 6
Phones, *see* Telephones
PIM (port interface module), 26

POTS (plain old telephone) sets, 33; *see
also* Analog
Predictive dialing systems, 97–99
Prefix codes, 3–4
Preparedness reviews, pre-
implementation, 139
Priority calling, 253
PRI (primary rate interface), 14
Privacy issues, 170–171
Private branch exchanges, *see* PBX
Private dialing plans, least cost routing
by, 4
Processors, speed of, 25
Project teams, implementation,
129–131
Pronexus VBVoice, 44–48
Proposals, *see* RFP
PSTN (public network), 4, 7, 8
Pulse code modulation (PCM), 10, 12

Q

QOS (quality of service) protocols, 147,
156
QSIG Protocol, 23–24
QualityCall agent evaluation system,
104–105
Queue time, 101

R

Ramp-up timing, 194–195
RBOC (regional Bell Operating
Companies), 2
RBS (robbed bit signaling), 14
Recorded announcements, 20
Rectifier, 16
Redirected calls, off-net enabled, 150
Redirect notifications, 252
Reduced instruction set (RISC) chips, 25
Redundancy, factors of, 25–26
Remedy Action Request System, 71–72
Remedy Help Desk, 71–81
Rep dials, 20
Resource Reservation Protocol
(RSVP), 159
RFP (request for proposal)
consultant's role in, 120–121
employee input on, 120
financial analysis of, 124–126

format of, 121
response evaluation, 121–126
vs. RFQ, 117–118
sample of, 215–248
selection committee, techniques
    for, 118–120
site visits for, 118–120
and updating existing systems, 121
vendor, evaluation of, 122–124
RFQ (request for quotation), 117–118
Right to privacy issues, 170–171
Ringing patterns, 252
RISC (reduced instruction set) chips, 25
Riser cables, 174
Robbed bit signaling (RBS), 14
Rotary dialing, 6–7
Route selection, automatic (ARS), 137,
    253
RSVP (Resource Reservation Protocol),
    159

S

S.100 standard, 68
Satellite communications (VSAT), as
    wireless option, 9
Scrambling devices, 185–186
Screen dumps, in SMIO switches, 16
Script Builder, 36
Seagate Software, Inc., 272–275
Security
    access services for, 188–189
    confidential information and,
      185–187
    digital cellular and, 187
    enterprisewide dialing plan, 187
    features checklist for, 30–31
    IP encryption for, 186
    malicious pranks and, 187–188
    scrambling devices for, 185–186
    threats to, 179–180
    toll fraud and, 180–185
    voice mail and, 188
    voice verification and, 180, 183,
      188, 189
Security violation logins, 252
Selection committee, RFP, 118–120
Self-excited linear prediction (VSELP),
    12

Servers
    for call accounting, 166–167
    client/server telephony, 66–67,
      71–73
    communication server networking,
      22
    Definity G3R, 249–253
    TSAPI (telephony server API),
      61–62, 65
Service access codes, 3
Service-level agreements (SLA),
    see SLA
Service objective percentages, 101
Service observation warnings, 251
Service providers, *see also* Carriers
    outsourcing by, 196–199, 205–209
    outsourcing call centers, 212–213
    service-level agreements with,
      209–213, 245–260
Shadow memory, 26
Siemen Phonemail, 6
Simple Mail Transfer Protocol (SMTP),
    93
Simple Network Management Protocol
    (SNMP), 166
Site visits, RFP, 118–120
Skills-based routing, 20
SLA (service-level agreements)
    carrier monitoring and, 199–201
    carriers, specifications of, 201
    checklist for, 209–213
    rate negotiations with carriers,
      191–196, 197–198
    sample of, 255–260
SMIO switches, screen dumps in, 16
SMTP (Simple Mail Transfer Protocol),
    vs. VPIM, 93
SNMP (Simple Network Management
    Protocol), 166
Software, *see also* Applications
    for call center management, 100,
      104, 261–275
    component, 66
    for help desks, 85
    IVR development software, 35–38
    middleware, 70–81
    networking capabilities and, 22
    for predictive dialing systems, 98–99
    for realtime events, 66